SpringerBriefs in Intelligent Systems

Artificial Intelligence, Multiagent Systems, and Cognitive Robotics

Series Editors

Gerhard Weiss, Maastricht University, Maastricht, The Netherlands
Karl Tuyls, University of Liverpool, Liverpool, UK; Google DeepMind, London, UK

This series covers the entire research and application spectrum of intelligent systems, including artificial intelligence, multiagent systems, and cognitive robotics. Typical texts for publication in the series include, but are not limited to, state-of-the-art reviews, tutorials, summaries, introductions, surveys, and in-depth case and application studies of established or emerging fields and topics in the realm of computational intelligent systems. Essays exploring philosophical and societal issues raised by intelligent systems are also very welcome.

More information about this series at http://www.springer.com/series/11845

Rick Evertsz · John Thangarajah ·
Thanh Ly

Practical Modelling of
Dynamic Decision Making

 Springer

Rick Evertsz
School of Science
RMIT University
Melbourne, VIC, Australia

John Thangarajah
School of Science
RMIT University
Melbourne, VIC, Australia

Thanh Ly
HMAS Stirling
Defence Science and Technology Group
Rockingham, WA, Australia

ISSN 2196-548X ISSN 2196-5498 (electronic)
SpringerBriefs in Intelligent Systems
ISBN 978-3-319-95194-2 ISBN 978-3-319-95195-9 (eBook)
https://doi.org/10.1007/978-3-319-95195-9

Library of Congress Control Number: 2019935986

This Springer imprint is published by the registered company Springer Nature Switzerland AG
The registered company address is: Gewerbestrasse 11, 6330 Cham, Switzerland

Acknowledgements

We would like to thank the Defence Science and Technology (DST) Group for their support of this work, and their ongoing invaluable discussions of important research questions in the modelling of tactics.

We would also like to thank the Defence Science Institute (DSI) for their encouragement and sponsorship of this work.

Contents

Acronyms

AOSE   Agent Oriented Software Engineering
ASW   Anti-Submarine Warfare
BDI   Beliefs Desires Intentions
ESM   Electronic Surveillance Measures
HVU   High Value Unit
L1-SA   Level 1 Situation Awareness (Perception)
L2-SA   Level 2 Situation Awareness (Comprehension)
L3-SA   Level 3 Situation Awareness (Projection)
MOE   Measure of Effectiveness
MOP   Measure of Performance
OODA   Observe Orient Decide Act
PSM   Problem-Solving Method
ROE   Rules of Engagement
RPD   Recognition-Primed Decision Making
SA   Situation Awareness
SHOR   Stimulus Hypothesis Options Response
STG   Surface Task Group
TDF   Tactics Development Framework
TMA   Target Motion Analysis
UAS   Unmanned Aerial System
UxV   Unmanned Vehicle

Chapter 1
Why Model Dynamic Decision Making?

A supply ship with two escorts is travelling through a conflict zone; the ships have received intelligence that there may be an enemy submarine ahead; consequently, they adopt a defensive posture, with the two escort ships out in front, in case they have to intercept the submarine. The submarine has detected their approach and calculates a rendezvous point from which to attack the supply ship. To avoid the two escorts, it tries to flank them so that it can fire a torpedo at the supply ship (Figure 1.1). Suddenly, it notices that one of the escort ships (Escort 1) has turned to port and appears to be on an intercept course. From a strategic perspective, the submarine is far more valuable than the supply ship, and so the submarine commander decides to abort the attack and flee.

This is one possible outcome. Many tactical variations are feasible; for example, instead of flanking, the submarine could have continued on its course, descended to hide on the ocean floor, and then attacked the supply ship after the escorts had passed overhead. The choice of tactic is situation dependent; for example, hiding on the ocean floor is only viable if it is not too deep, and the terrain is uneven, allowing the submarine to conceal itself within the ocean floor's contours.

Not only can tactics vary; in the real world, one has to be aware of the *dynamics of the situation* and, if necessary, drop one's current course of action to deal with the new situation. One of the escorts did this when it decided to temporarily abandon its goal of escorting the supply ship and adopted a new course of action, namely, to pursue the submarine. Similarly, the submarine dropped its goal of attacking the supply ship, and decided to flee, rather than risk being fired upon by the pursuing escort ship.

Many tactical situations involve a *team* of individuals working together to achieve their joint objective. For example, the two escorts should ideally work as a team to achieve their joint goal of protecting the supply ship from attack. Rather than just having the nearest escort turn to intercept the submarine, its teammate could change tack and move to form a defensive shield between the submarine and the supply ship.

This submarine scenario hints at a large class of problems where decisions have to be made under changing conditions. However, there are also many cases where

© Springer Nature Switzerland AG 2019
R. Evertsz et al., *Practical Modelling of Dynamic
Decision Making*, SpringerBriefs in Intelligent Systems,
https://doi.org/10.1007/978-3-319-95195-9_1

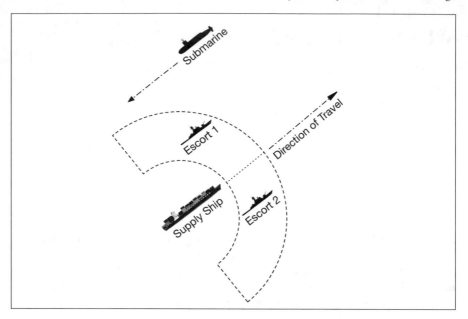

Fig. 1.1 Submarine tries to flank ships

decisions can be made under the assumption that the situation is fairly static. For example, when a doctor sits down with a patient to diagnose an illness, it is very unlikely that the patient's symptoms will change during the course of the consultation. Although such static problems are by no means trivial, decision making in dynamic scenarios generally requires a different emphasis, because of the need to adapt one's course of action when the changing situation means that it is no longer effective.

This book is concerned with how to model the dynamic decision-making capabilities of individuals as well as teams, with a view to *engineering* competent, autonomous decision-making systems for applications such as **UASs** (Unmanned Aerial Systems) as well as simulations that implement virtual human characters. In the autonomous UAS case, the UAS should be engineered to have decision-making capabilities comparable to those of human pilots so that it can successfully achieve goals such as dealing with bad weather, avoiding hostile threats, and working collaboratively with human team members. In the simulation case, we want the virtual human characters to behave realistically for applications such as disaster management training, and wargaming. These applications all pertain to *dynamic* environments.

According to Brehmer [9], dynamic decision making relates to:

> ... conditions which require a series of decisions, where the decisions are not independent, where the state of the world changes, both autonomously and as a consequence of the decision maker's actions, and where the decisions have to be made in real time.

Dynamic, real-world decision-making problems tend to involve changeable and sometimes misleading information about the situation – information that may be recognised as invalid once clarifying input is acquired or when the situation itself changes significantly. For example, because of the limitations of passive sonar in submarines, information about the tactical situation is typically time-consuming to acquire, full of gaps, and is sometimes uncertain. This means that the submarine commander's assessment of the situation can oscillate between various hypotheses as seemingly contradictory data come in. This severely impedes the decision-making process, and much of the commander's time is spent building situation awareness whilst not being detected.

For such problems, decision making tends to be time pressured, highly context sensitive, and constantly open to revision. This need to juggle goals and priorities means that standard software engineering approaches to modelling are a poor fit because they do not explicitly represent the key aspects of decision making, such as goal-based reasoning and the ability to drop a plan of action if it is unlikely to achieve the current goal.

In general, before one can model decision-making, one has to elicit the details from human experts. This turns out to be a very challenging endeavour because much of human expertise is tacit, and so it is difficult for domain experts to introspect about why they make the decisions they do. This so-called *knowledge acquisition bottleneck* was recognised very early on as a key problem in the construction of intelligent systems [28]. Pioneering work on the knowledge acquisition problem was performed predominantly by those developing **expert systems** - computer models that were intended to be comparable in domain-specific competence to human experts. However, these approaches fall short when it comes to eliciting dynamic decision-making expertise because they do not focus on how decision making in dynamic environments requires one to notice change, determine whether the current course of action is still effective, and, if it is not, choose an alternative approach.

To address the gaps in the elicitation and modelling of dynamic decision-making behaviour, we took Prometheus [49], a prominent **AOSE** (Agent Oriented Software Engineering) methodology, and extended it to create **TDF** (Tactics Development Framework) [27, 26, 24, 25]. In the past, a major problem for modellers was the lack of a common language for the specification of dynamic decision making. The development of such models usually requires collaboration between several communities with different skill sets and perspectives, for example, domain experts, operations analysts and software engineers. Within a given project, each community tends to use a different approach. For example, domain experts typically rely on natural language and informal diagrams, whereas software engineers might use UML [35] or go straight to program code. This makes communication between those communities difficult, and the links between the products of each community are tenuous and easily lost as the modelling process moves from domain experts through to software engineers.

With TDF, our objective has been to develop a methodology and modelling representation that is (i) easy for domain experts to understand, (ii) straightforward for operations analysts to apply, and (iii) for software engineers to use, debug and maintain. We want the representation to be easy to understand because this will help with knowledge elicitation and model validation. If domain experts can understand the representation, then it gives them a sense of ownership over the unfolding models, allowing them to relate the subsequent behaviour back to the models so that they can directly critique any shortcomings. Consequently, TDF offers diagrammatic views of decision-making models, rather than textual ones.

The goal of the TDF methodology is to facilitate the engineering of dynamic decision-making systems. To this end, it supports the elicitation of expertise from human decision makers, and the specification and design of **agents** [68] that will embody the elicited decision-making requirements. Agents are a widely used abstraction for encapsulating and expressing the behaviour of decision-making entities; for example, an agent can be used to represent a submarine commander or an autonomous UAS. Effective tactical performance requires capabilities such as the balancing of reactivity with proactivity. The tactical decision maker must be goal-directed, able to change tack when the situation shifts significantly, and able to coordinate its activities with peers who are working towards the same goal. It has been argued that these capabilities, namely autonomy, reactivity, proactivity and social ability, are characteristic of agent-based systems [68], and that they are not well supported by most programming paradigms. With this in mind, TDF is based on the agent-oriented Prometheus [49] methodology. Prometheus and other AOSE methodologies are concerned with how to specify, design, implement, validate and maintain agent-based systems. However, from the perspective of modelling dynamic decision making, no single AOSE methodology collectively tackles knowledge elicitation, dynamic goal-oriented control structures, team modelling, or the expression of tactics at a high level of abstraction. TDF is the first methodology to focus on the full development cycle required to model dynamic decision making, from knowledge elicitation through to implementation. Note that, although TDF is primarily focused on the elicitation and modelling of **tactics**, it can also be applied to the modelling of more general behaviour that is not necessarily tactical.

So, what do we mean by **tactic**? Military science definitions of the term, tactic, tend to emphasise the adversarial aspect. More generally, we view tactics as:

> ...the means of achieving an immediate objective in a manner that can be adapted to handle unexpected changes in the situation.

Tactics are distinguished from **strategy** in that the latter is concerned with a general approach to the problem, rather than the *specific* means of achieving a more short-term goal in a dynamic environment. A submarine commander's use of stealth to approach a target is a strategy, whereas the particular method used, for example hiding in the adversary's baffles (a blind spot), is a tactic. Tactics are concerned with the current, unfolding situation – that is, how to deflect threats and exploit opportunities to achieve one's objective. This view of tactics, as the means of achieving a short-term goal in a manner that can respond to unexpected change, seems to be common to all definitions, whether in military science, game theory, or management science. Clearly, there is a very large overlap between the meaning of **tactics** and **dynamic decision making**, and so we will use the two terms interchangeably in this book.

In theory, tactics can be automatically generated if the domain can be sufficiently well formalised. However, in complex, real-world domains this is currently not a practical approach, and so the dominant approach to tactics modelling has been to study domain experts and their decision making, and this **knowledge elicitation** aspect forms an important part of the TDF methodology. Having said that, TDF can be more generally applied to the creation of models that are not necessarily analogues of human decision making; there is nothing in TDF that specifically ties it to modelling human behaviour, it is just that, currently, the most sophisticated tacticians available are human and so TDF includes a process for eliciting their expertise. With this in mind, the next section outlines some of the background to modelling dynamic human decision making.

1.1 Dynamic Human Decision Making

Human decision making has been studied from many perspectives, including psychology, social psychology, behavioural economics, and more recently, cognitive neuroscience. The consensus is that human decision making is not exclusively rational; subconscious biases come into play, triggered by context or modulated by emotion. However, despite (or perhaps, because of) such biases, human experts seem to do well in dynamic, time-pressured environments [39].

We will leave the modelling of such subconscious factors to those who are interested in the fine-grained detail of human decision making. We are more concerned with the *engineering aspects* and so will focus on how to model the high-level, cognitive components of decision making, rather than their subconscious, moderating influences[1]. A number of existing models of decision making are couched at this higher, cognitive level; two of the most well known are Klein's **RPD** (Recognition-Primed Decision Making) model [39], and Boyd's **OODA** (Observe, Orient, Decide, Act) loop [16].

Given that the military routinely has to deal with very dynamic situations, which involve a hostile enemy and unpredictable outcomes, it is hardly surprising that many models of military command decision making have been developed over the years (see [18] for a review). Interestingly, they tend to be quite similar at the macro level, in that they structure tactical decision making into the same four sequential phases, although they use different terminology for the phases:

- **Perception**. Observing what is happening in the environment.
- **Comprehension**. Understanding how the observations relate to one's current goals including, perhaps, projecting into the future to divine how the situation might unfold.
- **Decision**. Given what one has observed, what one expects to happen, and what one's goals are, decide whether to continue with one's current course of action or pick a new one.

---

[1] Note that models, like those created in TDF, can incorporate such decision-making biases by targeting a suitable behaviour moderation architecture, such as PMFServ [56], that allows decisions to be influenced by emotional and corporeal factors such as fear and fatigue.

- **Action**. Perform the next step in one's chosen course of action.

Examples of models that fit within the aforementioned four-step decision-making loop include: (i) Boyd's OODA loop (though not published until after his death, elements of it were present in an unpublished essay as early as 1976 [51]); (ii) Wohl's SHOR (Stimulus Hypothesis Options Response) model [67]; (iii) Lawson's command and control model, comprising Sense, Process, Compare, Decide, Act [40]; (iv) the 4-Box model [36]; and (v) the RPD model.

The RPD model is interesting because it has been applied to a wide range of decision-making problems, and it reportedly accounted for 80-95% of the performance of naval surface ship commanders, tank platoon leaders and infantry officers studied [39]. There are a number of variants of the RPD model, but Figure 1.2 illustrates its key features and how, despite its apparent complexity, it maps naturally to the four-step schema outlined above.

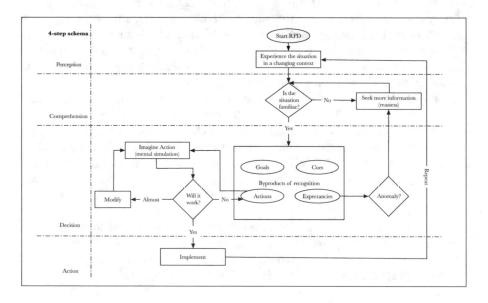

Fig. 1.2 Relationship between the RPD model and the four-step decision-making schema

The RPD process begins with the decision maker attempting to recognise the situation, seeking more information until the situation is recognised. The decision maker understands which percepts (cues) are key to the situation and adopts goals to be achieved as part of the decision-making process. Based on prior experience, there is an expectation regarding the events that should occur as the situation unfolds, and the decision maker also has a repertoire of effective actions. If the chosen action does not quite fit the current situation, then the decision-maker employs mental simulation to explore the consequences of modifying the actions to suit this particular case.

Interestingly, all of these perspectives on human decision making mesh neatly with an even simpler conception of human cognition: the **BDI** (Belief, Desire, Intention) model [52], which is described in Section 1.3. Although the BDI model derives from a very different school of thought, namely philosophy, its potential application to the modelling of human tactical decision making was recognised early on, for example in the SWARMM air combat model [46], and it forms the basis of TDF.

1.2 Situation Awareness

It is generally agreed that **SA** (Situation Awareness) is a key component of decision making; in fact, Boyd [16] argued that the 14:1 kill ratio of American vs. North Korean pilots in the Korean War came down to the better situation awareness afforded by the 360 degree view of their F-86's bubble canopy vs. the MiG-15's restricted rear view. In the military domain, SA has been defined as follows:

> ...the ability to have accurate real-time information of friendly, enemy, neutral, and non-combatant locations; a common, relevant picture of the battlefield scaled to specific levels of interest and special needs... [63, p. Glossary-7]

The most widely used definition of SA was coined by Mica Endsley:

> Situation awareness is the perception of elements in the environment within a volume of time and space, the comprehension of their meaning and the projection of their status in the near future. [21, p. 97]

These two definitions are compatible with one another, as are the many other definitions used in the literature. The first definition sees SA as an aspect of *mental state*; specifically, an individual's (or team's) task-relevant understanding of the current situation. In contrast, Endsley's definition is cast in terms of the *process* that leads to situation awareness; in the literature, this process is generally referred to as **situation assessment**. Endsley went on to develop a model of the situation assessment process that comprises three sequential steps:

- **Level 1 Perception (L1-SA).** This involves the acquisition of environmental data that is relevant to the task at hand. For example, when landing, a pilot will focus attention on the runway and will perceive attributes such as the rate of descent and how close the aircraft is to the ground. Irrelevant percepts are less likely to be attended to, such as whether the surrounding grass is lush or parched. In a task-oriented environment, perception tends to be focused on acquiring pertinent situational information. Thus, situation awareness is a relative term; the condition of the grass has no relevance to the task of landing an aircraft.
- **Level 2 Comprehension (L2-SA).** This concerns the further processing of L1-SA and inferring its implications for the task at hand. This inference process is

primarily deductive[2] but can also be abductive[3]. If on landing the aircraft the pilot observes that it is raining heavily, the runway is wet and there is a large patch of oil ahead, he can deduce that the aircraft's grip will be reduced. He might also *abductively* infer that an earlier aircraft had a mechanical fault leading to the loss of a large quantity of hydraulic fluid.

- **Level 3 Projection (L3-SA).** This is the most complex aspect of situation awareness. It uses the results of L1-SA and L2-SA to predict what will happen in the near term. In an adversarial situation, projecting into the future will involve reasoning about the goals and actions of the opponent and mentally modelling how the future might play out.

The Endsley model is qualitative in nature and the classification of knowledge into the three levels is sometimes subjective. For some decision makers, discerning a given attribute of the environment may involve inference (i.e. L2-SA) but for others with more experience, that attribute will be perceived without the need for explicit reasoning. In this latter case, the person will have experienced the situation so many times that its recognition becomes virtually instantaneous. To illustrate, a chess master will instantly recognise a situation in which a queen is pinned in front of the king; the queen cannot be moved without putting the king in check. In contrast, a complete novice would need to think about the situation before drawing the same conclusion. For the novice, all three Endsley levels might be involved: perceiving the individual pieces and their positions, comprehending the interrelationships between the pieces and projecting into the future (i.e. realising that moving the queen puts the king in check). In contrast, the ability of chess masters to perceive complex board positions is well documented, e.g. [12], and it is not unlikely that only L1-SA would be involved in perceiving the pinned state of the queen.

From this discussion of SA, it is clear that decision making is about more than just deciding on what course of action to adopt; a significant portion of a decision maker's cognitive effort will be devoted to situation assessment and determining how the current situation impacts the current goal. This has important implications for the computational modelling of dynamic decision making and has influenced the design of TDF.

1.3 BDI and the Theory of Mind

It is generally accepted that human beings employ a **theory of mind** to explain and predict the actions of others, as well as to provide a narrative for their own reasoning processes. According to the **intentional stance** [19], we attribute intentions to ourselves and others, and believe them to arise from a combination of our beliefs and

---

[2] In deductive reasoning, the available evidence is used to draw an irrefutable conclusion. If the premises are true, the conclusion *must* be true.

[3] In abductive reasoning, the evidence is used to develop a hypothesis that *explains* the evidence. However, it does not allow one to be certain about one's hypothesis; other hypotheses are possible.

desires. Through the work of Bratman [7] on practical reasoning, this philosophical perspective gave rise to the BDI (Belief, Desire, Intention) paradigm, a modelling approach that has been popular for many years in the multi-agent systems community. It has been claimed that the BDI paradigm is a good **folk psychology** model of human reasoning, i.e. a model of how we think we think [47].

The BDI model is a particularly parsimonious conception of rational agency, characterising agents as reasoning in terms of their **beliefs** about the world, **desires** that they would like to achieve and the **intentions** that they are committed to. Apart from its intuitive appeal to domain experts, it is a powerful computational abstraction for building sophisticated, goal-directed and reactive reasoning systems, and consequently is well suited to modelling decision making in dynamic environments.

Put very succinctly, a BDI agent performs a continuous loop in which it updates its beliefs to reflect the current state of the world, deliberates about what to achieve next, finds a plan for doing so, and executes the next step in that plan. Each time around this cycle it effectively reconsiders its options, yielding goal-oriented behaviour that is also responsive to environmental change.

Our experience in modelling dynamic decision making, in domains including air combat, infantry tactics, air traffic flow management and undersea warfare, suggests that modelling success is greatly influenced by whether domain experts find the conceptual framework in which the knowledge is cast to be intuitive. Domain experts are better able to understand and critique tactical models if those models correspond to how the experts *think they think*, and this improved understanding leads to better models. This was one of the main reasons for basing TDF on the BDI model of agency.

1.4 Team Oriented Decision Making

Work on the computational modelling of teams spans several decades. A **team** is a group of entities formed to achieve an explicit **joint goal**; it embodies structural relationships between entities predicated on the joint goal that all of the team members work towards.

At the heart of most approaches to team modelling is the consensus that **team mental state**, no matter how it is modelled, is essential for teamwork. Most studies of the mental attributes of teams have been either philosophical [8, 54] or theoretical [14]. From a philosophical perspective, there has been considerable debate on whether a team should be modelled as a separate entity or merely an aggregation of the different (mental) attributes of the agents that constitute it. Bratman [8] argued for the latter view, whereas Searle [54] insisted that there has to be such a thing as **collective intention** if a team is to perform adequately. Searle saw collective intention as a biologically primitive phenomenon, a prerequisite if you like, as indicated by the fact that one can ask a human team member something like "What is the intention of you and your colleagues?", and there will usually be a response of the form "We intend to...". He argued that individual intention and mutual belief are insufficient constructs on their own to support collective intention. Looking at how the field of team-oriented programming has developed, it seems that the majority of researchers and practitioners either agree with this assertion, or at least find it to be conceptually the most natural and efficacious of the two.

Leaving those arguments aside, we can agree that a team is an organisational structure that is formed to achieve a joint goal. Effective team performance relies on having a mutual understanding of what that goal is, having a shared understanding of the current situation, and having mechanisms for coordinating the activities of the team members. Team behaviour is a form of **joint activity** [29] that involves collaboration. However, joint activity can involve various types of mutual behaviour [6]; in particular (i) **co-allocation** – where two entities agree on and schedule their use of a resource they both need to access; (ii) **cooperation** – where two entities help each other with their different goals; and (iii) **collaboration** – where there is a shared goal that the entities are working towards. These three aspects of joint activity need to be considered when modelling team tactics. All teams involve some degree of coordination; however, collaboration is probably the most sophisticated of the three. Whereas co-allocation can occur without the entities knowing anything about each other's intentions, for all but the most trivial problems, collaboration in the real world requires the communication and mutual understanding of higher level cognitive constructs.

1.5 Agent Transparency in Human/Agent Teams

With the recent advances in the capabilities of autonomous **UxVs** (Unmanned Vehicles), human/machine teaming has become a very active research area, e.g. [65, 13, 44, 53]. The goal of such research is to enable more effective communication and shared awareness between humans and machines so that they can work well together. To function effectively as a team, there has to be mutual understanding and trust between team members; the human must comprehend *what* the artificial team members are doing and *why* they are behaving as they are. By "what", we mean more than merely observing what the team member is doing. Rather, we mean understanding what its *intention* is and how/why that intention relates to the overall objective of the human/artificial team, if at all. For example, a human team member, observing an autonomous photo-reconnaissance UAS descending unexpectedly, may want to know why the UAS has suddenly changed its course of action; perhaps its wings are icing up and it has decided to descend to warmer climes in order to avoid a potentially catastrophic stall. Thus, although descending may seem to have nothing to do with the overall photo-reconnaissance objective, it most certainly does in the context of wing icing; if the UAS stalls and crashes, it will not be able to fulfil its photo-reconnaissance role.

There are many ways for human team members to justify their trust in, or reliance on, an artificial agent's decision making. One basis for trust is that the agent has always succeeded in the past; after a while, the human can just rely on it, without having to know why it is behaving in a particular way. However, in the real world, it is risky to rely upon the omniscience of any agent, whether human or artificial. The artificial agent's reasoning may be completely sound, given the data it has about the current situation, but it may be missing a crucial piece of information that the human has. The danger here is that the human can place unwarranted trust in the artificial agent's decisions, and this can be just as dysfunctional as placing too little trust in the agent. The key requirement is for appropriately *calibrated trust*, that is, for the human to not be over or under reliant on the autonomous system. In the past, major accidents have happened because the human in the human/machine system had an inappropriate level of trust in the machine's performance, i.e. either too much, leading to an abrogation of responsibility during a critical incident, or too little, leading to the overriding of the machine's automated actions, with catastrophic consequences.

Previous research has shown that an effective means of developing calibrated trust is for the machine to expose its decision making so that the human can understand the underlying rationale [45, 41]. The degree to which the agent exposes its decision making is referred to as its level of *transparency*. Agent transparency can be fostered by effective communication with the human, and this can be mediated via the *tracing and monitoring* of the agent's computation as well as by the agent *explaining* its current and planned behaviour; cf. Debrief [38], VISTA [57], XAI [64] and TRACE [73].

Thus, agent transparency is a crucial component of trust; the human team member can establish the soundness of the autonomous agent's course of action if s/he can see what goal the agent is trying to achieve, what information it has based its decision on, whether that goal is a sub-goal of a higher-level goal, what other sub-goals the agent will need to achieve, what plans it has for achieving those goals, and so on. This means much more than being able to inspect the code that the agent is running and the current state of its data structures. The code and data structures are unlikely to be comprehensible to human team members, particularly in the context of a rapidly unfolding tactical situation. A key requirement of transparency is that the agent must communicate its decision making in a manner that can be easily and rapidly understood by the human. Thus, it should be compatible with how humans themselves *think they think*, for example, by being couched in terms of cognitive constructs such as *intention* and *belief*, in other words, Dennett's *intentional stance* [19], introduced in Section 1.3.

1.6 Summary

This chapter outlined the motivation for modelling dynamic decision making and introduced related background material on human decision making, situation awareness, the BDI model, team modelling, and the need for transparent agent models in mixed human/machine teams.

The modelling of dynamic decision making is important in a number of application areas including (i) capability analysis (simulation of tactical scenarios in order to identify gaps in capability); (ii) training (realistic simulation of non-player characters in a virtual environment); (iii) tactics development and evaluation (using decision-making models in a constructive simulation context to explore tactical options); (iv) autonomous systems (UxVs whose domain-specific decision-making competence needs to be on a par with that of humans); and (vi) human/machine teamwork (transparent agents that can explain their behaviour at the level of goals and intentions). In developing TDF, we were interested in those particular applications areas and so have focused on how to engineer effective and understandable dynamic decision-making systems, rather than emphasising other aspects such as the fine-grained modelling of human performance at the millisecond level, which is more of a concern for psychology research.

Following on from Dennett [19], TDF is grounded in the intentional stance and the BDI paradigm, as this has been shown to be a good folk psychology model of human reasoning (i.e. how we think we think), and so facilitates the development of decision-making systems that can be understood by domain experts and interacted with by human team members. Although TDF is rooted in the BDI model, it is agnostic with regard to the content of the decision-making models; one is free to overlay higher level frameworks such as RPD and the OODA loop.

1.7 Overview of Remaining Chapters

Chapter 2 introduces the artefacts and iconography that make up TDF's notation. An artefact is a *named* element of the model, and the concept is quite similar to that of a **class** in object-oriented design. Chapters 3 to 5 detail the three stages of the TDF methodology, namely the **Requirements**, **Architecture**, and **Behaviour** stages, and describe the methodological steps and diagrams of each stage. Chapter 6 presents an undersea warfare tutorial example of how to develop a model in TDF. In order to be practically useful, TDF requires tool support for creating and checking models as well as generating code templates from the models; the TDF Tool is introduced in Chapter 7. Finally, Chapter 8 wraps up the material presented in this book.

Chapter 2
Introduction to TDF

Chapter 1 introduced dynamic decision making and why it is important to be able to model it effectively; this chapter introduces TDF's diagrammatic notation. TDF is a methodology and tool that supports the design, development and deployment of sophisticated models of dynamic decision making. It provides a high-level, diagrammatic representation for modelling the dynamic aspects of tactics and has been developed to support the whole model development lifecycle, from knowledge elicitation, through model design, to the generation of platform-agnostic code templates. Its diagrammatic approach to tactics elicitation and specification is easy to understand and can be critiqued by domain experts – an important prerequisite for model validation. It provides traceability from requirements through to implementation and supports the construction of reusable and platform-independent tactics libraries.

TDF is based on over 25 years of experience in agent-based modelling of decision making and derives from the Prometheus methodology [49] and the requirements of analysts we have worked with, mainly in the military arena. The following main requirements have motivated the approach taken in TDF:

- **Tactics** – Support the grouping of behaviour into a tactic in order to facilitate its reuse, and to provide a high-level reference point for explaining agent and team behaviour. Tactics are goal-oriented, and, in working on a particular goal, the agent or team should restrict itself to the current tactic until the tactical goal is either achieved or abandoned, or the tactic fails to achieve the goal, in which case an alternate tactic should be tried if available. TDF supports this requirement through its provision of the **tactic** artefact (Section 2.20), **team tactic** artefact (Section 2.21), tactic diagram (Section 5.2), and team tactic diagram (Section 5.5).
- **Roles** – Facilitate the classification of the abstract functionality and responsibility that the entities in the system need to support. TDF supports this through its provision of the **role** artefact (Section 2.9), the relationships defined in the role enactment diagram (Section 4.1), role-based team structure definitions (Section 4.2), and role-based messaging and goal delegation in team plans (Figure 2.12).

© Springer Nature Switzerland AG 2019
R. Evertsz et al., *Practical Modelling of Dynamic
Decision Making*, SpringerBriefs in Intelligent Systems,
https://doi.org/10.1007/978-3-319-95195-9_2

- **Teams** – Explicitly represent teams in the system. A major shortcoming of earlier approaches has been that the designer had to find *implicit* ways to model teams. Team modelling should support the representation of the team's structure and joint goal (Figure 4.3), sub-teams (Figure 4.3), and the coordinated activity of team members (Figure 2.12).
- **Knowledge elicitation** – Support the elicitation of knowledge from domain experts by facilitating the diagrammatic representation of key scenarios (Section 2.4) and domain knowledge (Section 2.5).
- **Goal structures** – Provide a high-level, diagrammatic representation of how goals are decomposed into sub-goals. TDF provides the **strategy** diagram for this purpose (Section 3.3.2).
- **Reusable software components** – Provide a means of encapsulating functionality into components that can be reused across different **agent** types, or **team** types. TDF supports this via its **tactic** artefact (Section 2.20) and **team tactic** artefact (Section 2.21).
- **Agents** – Support the representation of individual behavioural entities as first-class elements of the model (Section 2.10).
- **Situation awareness** – Represent what the agent or team knows about the situation. TDF supports this by providing the **belief set** artefact (Section 2.15).
- **Environmental interaction** – Show what the agent can perceive in the environment and what actions it can perform. This is accomplished by showing **percepts** and **actions** in the agent diagram (Section 5.1). Overall environmental interaction of the system being modelled can be viewed in the I/O diagram (Figure 3.2).
- **Procedures** – Allow the specification of the procedures the agent/team follows to respond to events and achieve its goals. In TDF, agent procedures are expressed as **plans** (Section 2.17), and team procedures are expressed as **team plans** (Section 2.18).
- **Proactive and reactive behaviour** – Support the representation of proactive (goal-based) and reactive behaviour so that the agent or team is able to shift its focus when the situation changes significantly. In TDF, proactive behaviour is produced by agent **plans** and **team plans** that are invoked in response to the adoption of a goal. Reactive behaviour is produced by agent plans that respond to incoming **percepts**, **messages** and **belief set** updates.
- **Context** – Represent how context affects decision making and the choice of course of action. In TDF, contextual factors can be represented in a number of places: (i) as a **guard** on an arc in a strategy diagram (e.g. in Figure 2.9); (ii) as a context condition of a **plan** (Plan 1 in Figure 2.10), **team plan**, **tactic** (Section 2.20) or **team tactic**; and (iii) as a goal annotation in a strategy diagram (Figure 2.9).
- **Explainability** – Provide a representation that will facilitate the generation of explanations of agent and team behaviour. By virtue of its BDI roots, TDF encourages the development of explainable models of decision making. The grouping of behaviour into tactics provides a high-level basis for generating explanations. Furthermore, TDF's diagrammatic model representation offers a medium for intuitively presenting an agent's or team's decision making to the human observer.

2.1 Modelling in TDF

In keeping with its intended purpose as a practical methodology for modelling dynamic decision making, TDF addresses four main aspects of modelling:

- **Process.** Guidelines on the sequence of steps to be followed, what should happen in each step, and each step's purpose.
- **Artefacts and Relationships.** A TDF model comprises a number of **artefacts** and the relationships between them. Artefacts are the *named* elements that make up a model, for example, the agents, goals and tactics. A model should define the relationships between its artefacts, for example, the fact that an agent *has* goals that it can tackle and tactics it can *use* to achieve those goals.
- **Diagrams and Iconography.** The TDF methodology encourages the use of specific types of diagram that offer important views of the model. Each type of diagram can be constructed from a particular set of icons[1] (representing artefacts and nodes) and those icons can be connected together by arcs that show the relationships between them.
- **Route to Implementation.** It is all very well having a methodology for modelling decision making, but it is difficult to develop, verify and maintain the models without software tool support. Furthermore, when engineering applications, it is useful to automate aspects of the mapping from the model to executable code. These implementation aspects are discussed in Chapter 7.

2.2 TDF Stages

The TDF methodological process divides the specification and design of tactics into three main stages: **Requirements**, **Architecture**, and **Behaviour**, as outlined in the following three bullet points.

Note that, although the three stages tend to be performed in sequential order, in practice one can switch back and forth between stages.

- **Requirements.** This stage may involve knowledge elicitation from domain experts along with specification of system inputs and outputs, the main **goals** the decision-making system will have to handle, and what high-level **strategies** it will use to achieve those goals. During knowledge elicitation, **concept map** diagrams are used to informally express relationships between concepts, and **case study** diagrams show particular tactical scenarios that will be modelled during the design process. Before TDF, an analyst attempting to elicit and build a model

---

[1] Note that the icon's label is always positioned directly underneath the icon.

of tactical decision making would have to choose from a very wide and potentially confusing array of knowledge elicitation options, such as the many variations of Cognitive Task Analysis [17]. Our objective in developing TDF was to provide analysts with a straightforward and efficient methodology that prioritises the elicitation of the dynamic aspects of decision making over the more general aspects of knowledge elicitation. A key part of this is the provision of an intuitive diagrammatic representation of tactics that domain experts can relate to and critique during knowledge elicitation sessions, as well as later, during tactics evaluation and validation. We argue that this fosters, in domain experts, a sense of ownership of the emerging model, resulting in greater model validity, as well as traceability from the elicited models all the way through to the implementation and resulting behaviour.

- **Architecture.** Once the information has been elicited, it is time to specify and design the structure of the tactical decision-making system, namely the **roles** and the **agents** and **teams** that will enact the roles, as well as the interactions between the entities.
- **Behaviour.** Finally, the behaviour of the entities is specified. This comprises the **tactics** and **plans** that the entities use to achieve their goals.

2.3 Iconography

The following sections introduce the artefacts and iconography of TDF – its *language*, if you will. Each TDF artefact is denoted by an icon, and these icons are assembled into diagrams that express the relationships between the artefacts in a particular model. These diagrams will be introduced within the context of the overall TDF methodological process in Chapters 3 to 5, and a tutorial example is explored in Chapter 6 using an **ASW** (Anti-Submarine Warfare) scenario adapted from an agent-based simulation study of tactics for protecting an **HVU** (High Value Unit) [4].

Most TDF artefacts are atomic; in contrast, the case study, team and plan artefacts have a graph-like diagrammatic structure comprising artefacts and **nodes** linked together by **arcs**.

TDF objects can either be *named* or *unnamed*. Artefacts (Figure 2.1) are named objects that can be referenced in various parts of a model. That is, every instance of an artefact must be assigned a name that is unique for that artefact *type*. So, for example, a model can have a plan artefact named `Navigate to Destination` and a goal artefact of the same name; however, one cannot have two *different* plans named `Navigate to Destination` in the same model.

Nodes (Figure 2.2) are unnamed and denote logical relationships and particular types of computational step. For example, a **wait** node is used in a plan to represent a computational step that waits for a condition to become true before proceeding to the next step.

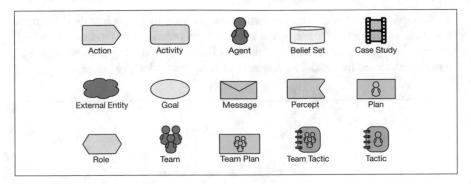

Fig. 2.1 Iconography of artefacts

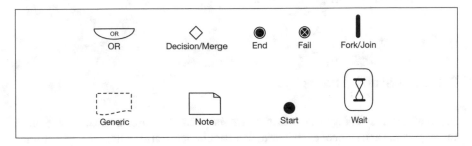

Fig. 2.2 Iconography of nodes

In a strategy diagram, a goal can be annotated to show the logical relationship between its sub-goals; the set of possible logical sub-goal operators is shown in Figure 2.3.

Fig. 2.3 Iconography of logical sub-goal operators

In a team plan diagram, a goal that is delegated to a role in the team can be annotated with the delegation operator 'ALL' to indicate that *all* of the role fillers must achieve the goal. Conversely, the 'ONE+' annotation is used to specify that *at least one* role filler must achieve the goal, although more could if they happen to achieve the goal at the same time (see the annotated arcs in Figure 2.4).

The use of artefacts, nodes, logical sub-goal operators and goal delegation operators should hopefully become clear as examples are presented in subsequent chapters. Each of the following sections introduces a different type of TDF object and, if relevant, provides a table of its properties and outlines its graph-like diagrammatic structure. To the extent possible, the artefacts are presented in the order in which they would typically be defined during the elicitation/modelling process, although in practice you are free to define artefacts in whatever order you prefer.

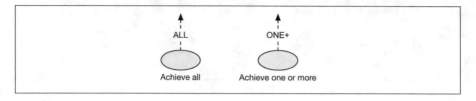

Fig. 2.4 Team plan goal delegation operators

2.4 Case Study Artefact

In Cognitive Task Analysis, the Critical Decision Method uses a case study to elicit the key events and *critical* decisions of a particular tactical situation; the case studies are documented informally, usually in prose. TDF takes this further by providing a diagrammatic means of representing case studies.

A TDF case study represents a linear sequence of the events that have occurred in a historical situation or could occur in a hypothetical one. However, in practice, the elicitation of a case study may bring up minor variations in how events can play out; it is vital to capture these variations whenever they arise during knowledge elicitation, otherwise they can be easily overlooked later on. For example, in a submarine case study, the domain expert might say something like:

> As ship's captain, I would instruct my navigator to set a *direct* course to the conflict zone, unless I had reason to suspect that there could be an enemy submarine around; in which case, I would adopt a random *zigzag* pattern to make it difficult for the submarine to compute a rendezvous point from which to attack us.

This *zigzag vs. direct* route decision is effectively a case study *variation*. In a similar vein, a submarine commander might explain:

> Close to the rendezvous point, I would give the order to prepare to fire a torpedo. If, during this period, incoming sonar is detected, and I estimate that the source ship is close enough to pick up the sonar returns, I would abort the attack and escape[2].

This too is a case study variation, but rather than involving an upfront choice between two courses of action, it involves aborting the current course of action

---

[2] Active sonar has to reflect off of the submarine's hull and be strong enough to return back to the sending ship. Thus, even though the submarine detects the incoming sonar pings, it doesn't follow that the ship will be able to detect their return.

(torpedo firing) in favour of another. TDF's diagrammatic case study formalism can support the expression of both types of variation. The upfront choice between two courses of action is represented by annotating the diagram with a **note** node, whereas aborting a course of action is represented as an **interruption** to an **activity**. The important point to grasp is that TDF allows the capturing of minor case study variations that would otherwise be lost; major variations, on the other hand, should be captured in separate diagrams.

Note that a TDF case study always proceeds forwards in time; it is not possible to represent the process-like idiom of looping back to an earlier part of the case study.

Property	Description
Name	A unique case study identifier.
Situation Description	Free form prose corresponding to what is termed a **vignette** in military parlance. Typically covering aspects such as the main objectives of the participants, the tactical situation (e.g. the fact that the RED force is concealed under tree cover), operational constraints (e.g. do not enter any no-fly zones), risks (e.g. RED force's strength is unknown), and opportunities (e.g. RED force is not expecting an approach from the south, so a surprise attack is an option for BLUE force).
Synopsis	A narrative of the events that occur in the case study.
Body	The graph-like diagrammatic structure of the case study.

Table 2.1 Case study properties

In addition to the first three properties shown in Table 2.1, a case study has a fairly rich diagrammatic structure, called its **body**, that is typically partitioned into **participants** (Section 2.4.3) and **episodes** (Section 2.4.2). The artefacts and nodes that make up a case study's diagrammatic structure are described in Section 2.4.1. The process for eliciting a case study is described in Section 3.2.1, and a detailed example is presented in Section 6.3.2.1.

2.4.1 Case Study Body – Artefacts and Nodes

- **Start Node.** Denotes the beginning of a participant's sequence of steps. Consequently, if a case study has a number of participants, it will have a number of start nodes (as illustrated in Figure 2.7).
- **Activity Artefact.** Encapsulates a sequence of one or more decision-making steps and actions. See Section 2.16 for more detail.

- **Fork/Join Node.** Used to represent the concurrent branches of activity of a participant. The fork node splits a branch into separate concurrent branches, and the join node is used to connect them back together once the concurrent activities are over.
- **Interruption Node.** Represents an unexpected interruption to an artefact or node triggered by a general condition such as 'IF: Incoming torpedo' or 'IF: 10 minutes have elapsed'. Figure 2.5 shows an excerpt of a case study where the 'Travel to conflict zone' activity is part of the normal left-to-right sequence of the case study (indicated here by the left and right ellipses). If there is an incoming torpedo, the activity fails to complete, and exits via the vertical arrow that is labelled with the condition 'IF: Incoming torpedo'. To view this icon within the context of a case study see Figure 2.6.
- **Message Artefact.** Used to represent communication between the participants in the case study.
- **Note Node.** Annotates elements of the case study for the purposes of documentation.
- **Percept Artefact.** Represents the important environmental events that a participant is perceiving at any given point in the case study. See Section 2.12 for further detail.
- **Wait Node.** Represents a point where a participant is waiting for a condition to become true. It can be used to express temporal information, such as waiting for an elapsed amount of time, or it can be used to represent a more general condition, such as waiting for an event to occur.
- **End Node.** Represents the end of a sequence of steps for a given participant.

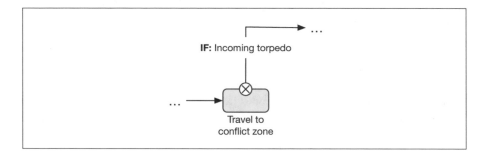

Fig. 2.5 Interruption of an activity

2.4.2 Case Study Body – Episode Column

Because case studies can be quite complex, TDF allows them to be partitioned into smaller, meaningful units called episodes. An episode is a named, encapsulated portion of the case study, represented as a vertical column. Figure 2.6 shows the `Travel and Handle Torpedo` episode of a case study for the `HVU` participant. Episodes can be shared across different case studies, which facilitates the specification of case study variations by using episodes to represent the common sub-sections of the case study variants.

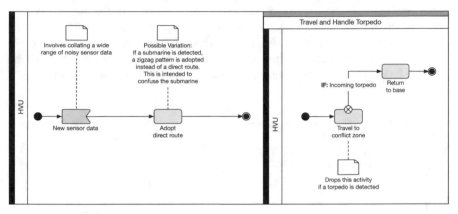

Fig. 2.6 HVU participant with episode

2.4.3 Case Study Body – Participant Row

At its simplest, a case study consists of a sequence of nodes and artefacts that represent how the case study unfolds over time. However, case studies usually involve a number of role players, whether human, machine or a team. Within a TDF case study, these role players are called **participants**, and TDF allows a case study to be partitioned into concurrent streams of activity corresponding to its participants. This makes multi-participant case studies easier to understand because one can see what each participant did, including how and when participants interacted with one another.

In a case study diagram, participants are represented as labelled horizontal rows (Figure 2.7). Although participants can be used to represent individuals, roles or teams, these concepts do not have any special status in a TDF case study. You can make a note of the participant type by giving it an appropriate name, such as `Escorting (role)`, but the `role` suffix is just documentation; TDF does not use the label to discriminate between different types of participant[3]. If the internal interactions of team members need to be documented, the team participant can be

---

[3] Note that later on, in the modelling stages, agents, teams and roles are explicitly modelled as first-class entities. This distinction is not made in case studies because the elicitation process can get bogged down when the analyst tries to decide whether a participant is a more general role or an

subdivided into the individual team members. A team is shown in Figure 2.7 where the 'Convoy' participant (a team) is partitioned into an 'HVU' participant (a role) and an 'Escort' participant (a team) comprising a helicopter and a ship.

Communication between two participants is represented by a message artefact, e.g. 'change velocity' (Figure 2.7). Communication is assumed to be instantaneous and so roughly represents time-based synchronisation between participants; hence, in Figure 2.7, the 'Assign sonar policy' activity of the 'HVU' overlaps with the 'Execute dipping sonar search pattern' activity of the 'Front Helicopter'. Synchronisation is also indicated by episodes because all of the events in an episode occur during the time span of the episode. Synchronisation can also be indicated by connecting artefacts across participants with an undirected dashed arc.

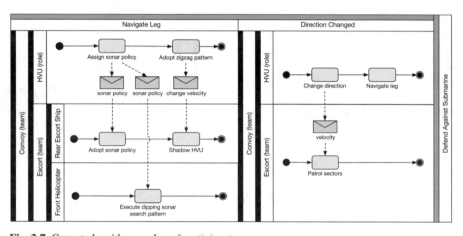

Fig. 2.7 Case study with a number of participants

Episodes also enable one to change the granularity of the view of groups of participants. For example, in Figure 2.7, the 'Navigate leg' episode shows the behaviour of the two participants in the 'Escort' team. However, in the adjacent 'Direction changed' episode, the 'Escort' team is collapsed so that the team members are no longer shown; this allows one to summarise the behaviour in that episode at the level of the team. The subsequent 'Defend Against Submarine' could collapse the view further to just show the activity of the 'Convoy' team, or could open it up again to show what the helicopter and escort ship are doing.

agent/team type. Although it is an important distinction during design, it has limited value during knowledge elicitation (at least compared to the overhead it introduces during an interview session).

2.5 Concept Map

The **concept map** is a diagrammatic representation of the domain, used as a focus
for discussion during knowledge elicitation. As shown in Table 2.2, it has a body –
a graph-like structure that informally represents the declarative information of the
domain in terms of labelled nodes (concepts) and arcs (relations between concepts).
One is free to create arbitrarily labelled nodes and arcs, and this makes concept maps
relatively quick to construct; an important attribute for any knowledge elicitation
formalism.

Property	Description
Name	A unique concept map identifier.
Body	The graph-like diagrammatic structure of the concept map, comprising generic nodes and any TDF artefact.

Table 2.2 Concept map properties

Initially, the concept map's body mostly consists of **generic** nodes (see Section 2.6), but as knowledge elicitation progresses, these tend to get converted into
one of the TDF artefacts (Figure 2.1). Figure 2.8 shows part of a concept map
that has been partially classified and so contains artefacts as well as one remaining generic node. Note that the arcs do not have to be labelled.

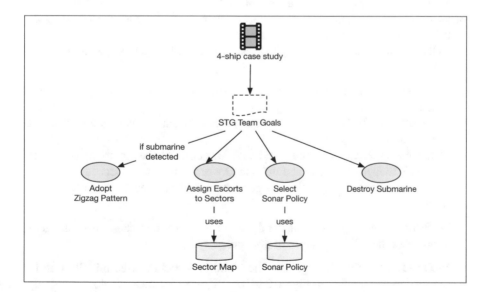

Fig. 2.8 A small part of an anti-submarine warfare concept map

2.6 Generic Node

The **generic** node is a general purpose element that is used to represent uncategorised nodes in the concept map (see Figure 2.8).

2.7 Goal Artefact

The goal artefact is typically used to represent something that an entity may need to achieve; for example, in Figure 2.8 'Destroy Submarine' is a goal that STG (Surface Task Group) needed to achieve during the '4-ship case study'. Goals are a very important aspect of dynamic decision-making models because such models are generally about *purposeful* behaviour, rather than mere reactive behaviour.

2.8 Strategy

Recall from Chapter 1 that, in contrast to tactics, strategies are concerned with general approaches to problems, rather than the specific details of the method used to solve the problem. In TDF, a **strategy** (see Table 2.3 for properties) is represented as a high-level goal tree, with a root goal that is termed its **objective** and child goals that are sub-goals, i.e. a decomposition that satisfies the objective. A strategy is *high-level* in the sense that the tree is typically only one or two levels deep. This is because, in practice, goal trees that are more than a couple of levels deep typically start to tackle the nitty-gritty detail of tactics.

TDF provides the following logical goal operators for combining sub-goals together:

- **AND.** Requires that *all* of the sub-goals be achieved; they are adopted one at a time in any order.
- **Sequential AND.** Like **AND** but requires that the sub-goals be adopted in left-to-right order.
- **OR.** Means that only one sub-goal has to be achieved (i.e. an *exclusive-or* relationship); they are tried one at a time, in any order, until one succeeds.
- **CON.** Specifies that *all* of the sub-goals be achieved; they are adopted concurrently.

To facilitate the expression of the dynamic variables that can affect a strategy, TDF provides the following types of goal property:

- **INITIAL CONTEXT.** Expresses the conditions that must be initially true for the goal to be adopted; in Figure 2.9, the 'Attack fighters' goal is only applicable if the ratio of enemy fighters to one's own team is less than two to one.

- **SUCCEED IF.** Specifies the conditions that result in the goal being achieved independently of the achievement of its sub-goals. This is illustrated in Figure 2.9 where the `Attack fighters` goal succeeds if the enemy retreats.
- **WHILE.** Expresses the conditions that have to remain true while the goal is being pursued. If the `WHILE` condition is no longer true, the associated goal is dropped, but in contrast to a **maintenance condition**, the system does not try to repair the situation. In Figure 2.9 the `Attack fighters` goal is pursued as long as the enemy fighters do not outnumber the team by a ratio of two to one, or more.

The concept of a **maintenance condition** differs from a `WHILE` condition in that it specifies that the system should take action to repair the condition that is no longer true. For example, if in Figure 2.9 the `WHILE` condition was a maintenance condition, then, while attacking the fighters, the system should *maintain* an enemy/self ratio that is less than 2/1. Maintenance conditions should be used sparingly as they can have very subtle consequences; it is incumbent on the designer to ensure that the system will adopt an appropriate repair method, which can be executed without compromising the achievement of the overall goal. Maintenance goals are not well suited to more complex cases where repair of the maintenance violation would involve significant time or the implementation of changes that endanger the achievement of the overall goal. For example, one might be tempted to add in the maintenance condition: `MAINTAIN enough fuel to return to base` so that the entity refuels if its fuel reserves are getting too low. However, breaking off an attack to refuel would take considerable time and would likely involve a significant change in location. It would not be practical to refuel and then pick up where you left off while attacking the enemy fighters. Because of these potential pitfalls, TDF does not provide a maintenance goal property.

Conditional statements can be used on arcs to specify the factors that constrain whether a sub-goal is adopted; this is shown in Figure 2.9, where the **guards** `outnumber enemy` and `NOT outnumber enemy` express which branch to take. The following keywords can be used in goal properties as well as within guards on an arc: `AND`, `OR` and `NOT`. As you will see later, these keywords can also be used in some other parts of the model, including **team structure** definitions, agent **plans**, and **team plans**. The meaning of the strategy in Figure 2.9 is: if you outnumber the enemy, adopt the goal of attacking line abreast; otherwise use a pincer formation, by first determining the flank assignments of your attacking aircraft, then flying to the attack position, and finally attacking the enemy.

Property	Description
Name	A unique strategy identifier.
Body	The strategy tree.

Table 2.3 Strategy properties

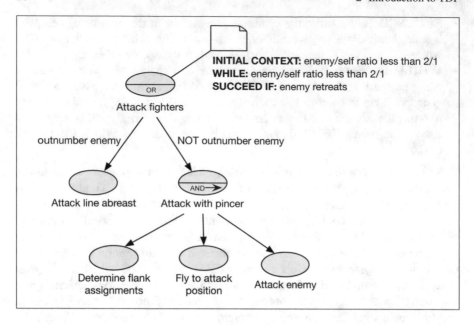

Fig. 2.9 Air combat strategy for attacking enemy fighters

2.9 Role Artefact

Goals represent what can be achieved by the system being modelled. In practice, goals are adopted by particular entities, whether individuals or teams. When modelling a system, it is useful to group *related* goals together – related in the sense that a particular entity tends to take on the responsibility for achieving that group of goals. In TDF, such a group of goals is represented as a **role** (see Table 2.4 for properties). An entity's role not only represents a *responsibility* that it has, but also *functionality* that it possesses; after all, for a system to function effectively, its members should only take on the responsibility for roles they are capable of fulfilling.

To illustrate, consider the roles required to tackle a forest fire. At headquarters, different people will take on different roles. For example, the **incident controller** role involves goals such as generating the battle plan and assigning strike teams to sectors. On the ground, a tanker team will comprise personnel with different roles, such as a **crew leader** who is responsible for the decision making of the overall team, and the **driver** who can handle goals such as safely negotiating intersections on the way to the fire.

Property	Description
Name	A unique role identifier.
Goals	The goals that the role can take on.

Table 2.4 Role properties

2.10 Agent Artefact

The **agent** artefact (see Table 2.5 for properties) is used to represent *individuals* that may enact roles. Note that a given agent does not have to enact any roles but may have multiple roles, for example, enacting the crew leader and driver roles in a fire engine team.

Property	Description
Name	A unique agent identifier.
Actions	That the agent can perform.
Belief Sets	That the agent can read from and/or write to.
Goals	That the agent can achieve.
Messages	Sent or received by the agent.
Percepts	Handled by the agent.
Plans	The agent has.
Roles	The agent can enact.
Tactics	The agent has.

Table 2.5 Agent properties

2.11 Team Artefact

The **team** artefact is used to represent collections of entities that work together towards a common goal. The entities can be agents or teams, and by having sub-teams, TDF supports the representation of team hierarchies of arbitrary depth.

A team has most of the properties of an agent, except for percepts and actions, and it substitutes team plans for agent plans, and team tactics for agent tactics.

2.12 Percept Artefact

The **percept** artefact is used to represent individual *items* that an agent perceives in its environment and corresponds to L1-SA situation awareness, as described in Section 1.2.

2.13 Action Artefact

The **action** artefact represents an *atomic* operation that the agent can perform in its environment. An action artefact is used when there is no need to break things down further; for example, for a pilot, lowering the landing gear could be represented as an action to be performed.

2.14 Message Artefact

The **message** artefact is used to represent communication between entities, for example, when a sonar operator informs the ship's captain that a submarine has been detected.

2.15 Belief Set Artefact

Belief sets are used by agents and teams to store the *declarative* information (beliefs) they need to maintain. Declarative beliefs are essentially the facts the agent/team believes in, that is, *what* it knows. An agent or team will typically have a number of belief sets that structure its beliefs into meaningful collections. For example, a navigation agent might have a belief set that holds the waypoints to the destination; a separate belief set might hold the waypoints it has already traversed.

In TDF, the internal structure of beliefs is not modelled; this is left to the programming stage that comes after the models are completed.

2.16 Activity Artefact

The **activity** artefact is used in case studies and plans to encapsulate a sequence of one or more decision-making steps and actions. The activity is effectively a *black box*, in the sense that its internal steps are not shown. Ideally, an activity should be given a name of the form *'verb noun'*, or if necessary, *'verb-phrase noun-phrase'*. For example, `extinguish fire` (verb = `extinguish`, noun = `fire`) or `travel to area of operation` (verb phrase = `travel to`, noun phrase = `area of operation`).

2.17 Plan Artefact

Agent **plans** specify how to achieve goals and respond to events. TDF provides a diagrammatic plan representation (the plan **body**) that facilitates the expression of the general steps of a procedure without getting bogged down in implementation detail (see Table 2.6 for properties). There is a long tradition of using diagrams to represent procedures, for example, flowcharts [32], Petri Nets [50], recursive transition networks (e.g. PRS [31]), UML activity diagrams [35], and BPMN [48]. Because UML is widely used for diagrammatic software specification, TDF uses it as the basis for its plan diagram representation, modifying it where necessary to better support the modelling of decision making, for example by including goal artefacts.

Property	Description
Name	A unique plan identifier.
Body	The graph-like diagrammatic structure of the plan.
Context Condition	The specific context in which the plan is applicable.
Invocation Condition	The artefact that triggers the plan (in BDI parlance, makes the plan **relevant**). A triggering artefact can be a goal, percept, message or belief set.
Success Condition	The conditions under which the plan succeeds.
While Condition	This condition must remain true while the plan is executing. If it becomes false at any point, the plan is dropped.

Table 2.6 Plan properties

Like case studies, TDF plans are drawn from left-to-right; also, TDF's plan diagram notation has a large overlap with that used for case studies because this reduces the amount of notation to be learnt. However, plan diagrams represent *general purpose* procedures that an agent can apply to a situation, whereas a case study diagram represents a *concrete* sequence of steps that occurred, or could occur, in a particular situation. Figures 2.10 and 2.11 show four plan diagrams contrived to show all of the applicable nodes and artefacts; consequently, they are unusually complex and dense.

2.17.1 Plan Artefacts and Nodes

- **Action Artefact.** As described in Section 2.13.
- **Activity Artefact.** As described in Section 2.16.
- **Belief Set Artefact.** As described in Section 2.15. A belief set update in a plan is represented as an incoming arc from the belief set to the start node of the plan[4]. A plan also shows the belief sets it accesses and/or updates. An update to the belief set is shown as an offshoot of an activity in the plan; `Plan 3` in Figure 2.10 updates the `Faults` belief set if a fault is found during the launch sequence[5].
- **Decision/Merge Node.** A decision node represents a conditional choice between options; each option is numbered to indicate the order in which the options are considered (see the decision node at the end of `Plan 1` in Figure 2.10). When paired with a decision node, a merge node reconnects the separate options back into a single path (see `Plan 3` in Figure 2.10). In this example, if the torpedo bay door is open, it gets closed; otherwise, nothing needs to be done. The two branches are merged and the plan goes on to the goal `Execute launch sequence`.
- **End Node.** Terminates the plan and signifies that the plan was successful. A plan can have multiple end nodes.
- **Fail Node.** Terminates the plan and signifies that the plan was unsuccessful. As is the case with end nodes, a plan can have multiple fail nodes.
- **Fork/Join Node.** A fork node is used to create concurrent computational branches. In `Plan 2`, the `Launch jammer` and `Launch decoy` activities are executed concurrently; the two branches come together with a merge node. The merge node will transition to its outgoing arc as soon as one of its incoming arcs finishes. This means that the plan does not wait for both launch activities to finish; it will continue to the next node as soon as either the jammer or the decoy is launched. The join node synchronises its incoming branches; all of the incoming branches must finish for the join to continue to its outgoing arc. So, in the example in Figure 2.10, `Plan 2` finishes once the `Turn, dive, accelerate` activity finishes *and* either the jammer or decoy is launched (or both).

  The pairing of a fork node with a merge node provides a means of expressing an *asynchronous goal*. Normally, the goal on a given branch of a plan is a sub-goal, and so the plan waits until the goal is achieved before proceeding to the next step on that branch. In contrast, an asynchronous goal will be considered separately from the plan; the plan does not wait to see whether it was achieved or not. Thus, an asynchronous goal is a goal that the agent would like to achieve, but the current plan does not depend on the successful achievement of that goal. This can be expressed in a TDF plan by pairing a fork with a merge node, as shown in Figure 2.11. Here, the submarine will behave as it did in `Plan 2` but in

---

[4] In Figure 2.10 `Plan 2` is triggered when the `Sonar detections` belief set has `incoming torpedo` added. Note that `incoming torpedo added` is an annotation, and like all other annotations in TDF, it is free-form text.

[5] The text `fault` is an annotation and so is free-form text. Coupled with the directed dashed arc connected to the belief set `Faults`, it indicates that a fault is added.

parallel will try to achieve the goal `Determine escape plan`. However, because of the merge node, as soon as it has completed the activity `Turn, dive, accelerate` and one or both of the activities `Launch jammer` and `Launch decoy`, it will move to the end node, i.e. without waiting for the achievement of the `Determine escape plan` goal. That goal can be managed by another plan, in a separate thread of activity.

- **Goal Artefact.** If the goal artefact precedes the plan's start node, then it represents the plan's invocation condition. Goals that come after the start node are adopted as part of the execution of the plan, e.g. `Close torpedo bay door` in `Plan 3`. A given goal must succeed before the plan can move to the next step. A goal can also be delegated to another entity so that, on the branch in question, the plan waits until the entity achieves the goal before proceeding to the next step on that branch[6].
- **Interruption Node.** Represents how to handle an interruption to an activity, goal or a wait node. When coupled with a goal, an interruption node can be used to represent the concept of a **preserve**[7] goal. This is shown in `Plan 3` where the `Execute launch sequence` goal is pursued as long as a fault is not found.
- **Message Artefact.** A message can invoke a plan, for example `Plan 3`; this plan also shows a message being sent to the role `<sender>`. In TDF, the role label `<sender>` denotes the agent or team that sent the triggering message. It is quite common to need to reply to a sender, and so TDF plans allow this to be represented using the `<sender>` role label. This also means that you are not allowed to create a role named `<sender>`.
- **Note Node.** For attaching documentation to elements of a plan.
- **Percept Artefact.** Typically used to represent a reactive plan, i.e. a plan that is triggered by an external event[8]. Can also be used in conjunction with a **wait** node in the body of a plan to represent waiting for a percept from the environment.
- **Start Node.** Plan diagrams have a unique start node triggered by either a goal, percept, message or belief set update.
- **Wait Node.** Waits for a condition to become true. Can be used for a general condition, or a temporal one, such as waiting for an elapsed amount of time.

A plan can have a context condition that represents the situations in which the plan is applicable. Figure 2.10 shows the context condition `INITIAL CONTEXT: NOT under threat` in `Plan 1`; the plan is applicable when the goal `Navigate waypoints` is adopted, but only if it is not `under threat`. Additionally, a

---

[6] An example of goal delegation is shown in Figure 6.26, which is a team plan rather than an agent one. Nevertheless, the layout is the same apart from the fact that, in an agent plan, the goal is delegated to either another agent or a team, whereas in a team plan, it is delegated to a role in the team.

[7] A goal that is pursued until the condition to be preserved is violated [37].

[8] In `Plan 2`, the triggering belief set update could have been replaced by the percept `Incoming torpedo`, if that was a more natural way to model how the plan is triggered. Generally, a belief set update is used if the plan needs to be triggered after the situation has been analysed and comprehended (in other words, after L2-SA; see Section 1.2).

'WHILE' condition is specified that indicates that the plan should be abandoned if the battery level is no longer above 20%.

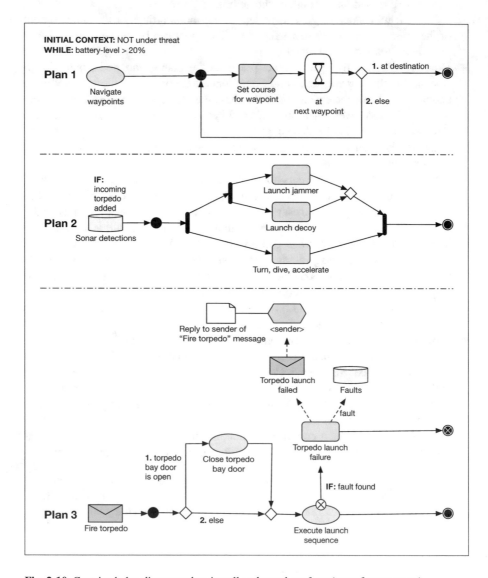

Fig. 2.10 Contrived plan diagrams showing all nodes and artefacts (apart from percept)

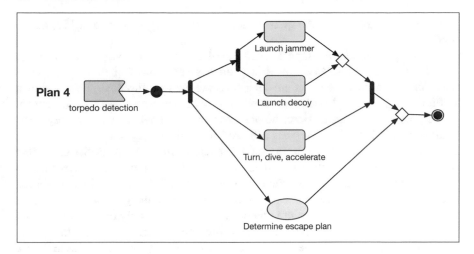

Fig. 2.11 How to express an asynchronous goal

2.18 Team Plan Artefact

As mentioned in Section 1.4, the notion of a *joint goal* is the key differentiator that distinguishes teams from other types of organisational structure. Team plans are used to specify how team members synchronise their approach to the joint team goal (see Table 2.7 for properties).

Property	Description
Name	A unique team plan identifier.
Context Condition	The specific context in which the team plan is applicable.
While Condition	This condition must remain true while the team plan is executing. If it becomes false at any point, the team plan is dropped.
Success Condition	The conditions under which the team plan succeeds.
Body	The graph-like diagrammatic structure of the team plan.

Table 2.7 Team plan properties

Team plans extend the agent plan diagram language with *role-based goal delegation* so that they can support flexible team coordination mechanisms; however, percepts and actions are disallowed because a team is an abstract entity that can only interact with the environment via one of its member agents. Each team plan is used to achieve a particular team goal, as illustrated in Figure 2.12 which handles the 'Pincer intercept' goal. Note that a team can have alternative team plans for a given goal, and this allows the specification of multiple approaches to achieving the goal.

A team plan *synchronously* delegates a goal to a role. By *synchronous* we mean that, before proceeding to the next step on a given branch of the team plan, it *waits* until the team members enacting the role achieve the goal. This is shown in Figure 2.12 – an illustrative but contrived 'Pincer Intercept' team plan for a 'Fighter Pair' team. Here, before adopting the goal 'Attack enemy', the team waits until the 'Leader' has achieved its goal to 'Fly to flank turn-in position'. As you will see in Section 4.2, a team structure can specify that a role requires more than one role filler; which means that if a team plan delegates a goal to a role, it will be passed to a number of role fillers. In such cases, the plan should specify whether it requires just one role filler to achieve the goal or all of the role fillers. In this example, the point is moot because there is only one 'Leader' and so the arc is labelled with the **delegation operator** 'ONE+', although the 'ALL' operator would have the same effect because there is only ever one 'Leader' in a 'Fighter Pair' team. An example of where the 'ONE+' operator should be applied is Search & Rescue, where the 'Search' role would be filled by a number of search teams, all looking for the same lost group of people. Here, we want the team to stop searching as soon as *one* of the teams finds the lost group. In contrast, before tackling a fire, a firefighting team should wait until *all* of the sub-teams arrive at the site; in this case the 'ALL' operator would reflect the intended behaviour.

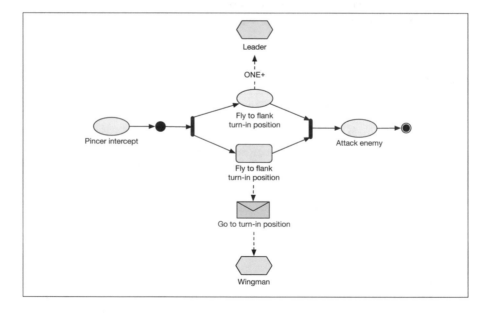

Fig. 2.12 Simplified pincer intercept team plan

2.18.1 Team Plan Artefacts and Nodes

- **Activity Artefact.** As per agent plan diagram.
- **Belief Set Artefact.** As per agent plan diagram.
- **Decision/Merge Node.** As per agent plan diagram.
- **End Node.** As per agent plan diagram.
- **Fail Node.** As per agent plan diagram.
- **Fork/Join Node.** As per agent plan diagram.
- **Goal Artefact.** As per agent plan diagram. Note that a goal can also be delegated to a role.
- **Interruption Node.** As per agent plan diagram.
- **Invocation Condition.** As per agent plan diagram but excludes percepts.
- **Message Artefact.** As per agent plan diagram; however, within a team plan, a message can be sent to a role.
- **Note Node.** As per agent plan diagram.
- **Role Artefact.** Represents the destination of a message or goal, as shown in Figure 2.12. The scope of role references is important in team plans. Messages or goals can target role fillers within the team but cannot reference roles outside of the team. Thus, if we assume that this team plan belongs to the 'Fighter Pair' team shown in Figure 4.3, then it cannot refer to the 'Strike' role filler of the 'Close Air Support' *ancestor* team. If the 'Fighter Pair' team needs to coordinate with the 'Strike' role filler, then this must be encoded in a team plan at the level of the 'Close Air Support' team. This constraint is imposed because the 'Fighter Pair' team cannot assume that it will always be a sub-team of a team that has a 'Strike' role filler.
- **Start Node.** As per agent plan diagram.
- **Wait Node.** As per agent plan diagram.

2.19 Interruption Node

The interruption node can be used in case studies, agent plans, and team plans, and has already been described in Sections 2.4.1 and 2.17.1.

2.20 Tactic Artefact

Earlier AOSE methodologies, such as Prometheus [49], offered a way of structuring agents in terms of discrete modules called **capabilities**. In such methodologies, a capability is an encapsulation mechanism that encourages good software engineering by allowing functionality to be aggregated and reused across agents. To facilitate modelling of dynamic decision making, TDF extends the capability concept and introduces the concept of a **tactic**. The tactic artefact fulfils two roles in TDF: (i) like a capability, it provides software engineering support by allowing behaviour to be encapsulated in a way that enables it to be reused to build solutions to similar problems; and (ii) it groups goal-based functionality into a tactic, so that the tactic can be used as a basis for explaining what the agent is doing at any given point in time (this requirement was discussed in Section 1.5).

To address reuse and sharing, TDF allows tactics to be annotated so that they can function as **design patterns**. Design patterns first became popular in object-oriented programming [30], but their use has also been investigated for building intelligent systems. A case for design patterns in human behaviour modelling is made in [58], and **generic tasks** [11] have been proposed as a means of representing high-level problem-solving in a way that captures the strategy using a vocabulary that is relevant to the task. In a similar vein, **PSMs** (Problem-Solving Methods) [43] express domain-independent, reusable strategies for solving certain classes of problem. A PSM comprises a definition of *what* it achieves, *how* to achieve it and what it *needs* to perform its function. In TDF, the *what* is expressed as the objective, the *how* as the plans, and the *needs* as the percepts, messages and belief sets.

A tactic can be decomposed into a hierarchy of sub-tactics, and this allows one to break down its functionality into smaller reusable components that can be composed together in different ways. To address the need to explain the agent's course of action to a human, when an agent adopts a tactic to achieve its objective, it will not use other ancillary methods that are not part of the tactic. In other words, the agent is *committing* to achieving its objective using *only* the tactic's plans and sub-tactics. If the agent cannot achieve its objective using the tactic, then it drops the tactic as a whole, and tries a different method if it has one (either another tactic or a plan).

Property	Description
Name	A unique tactic identifier.
Problem Description	A description of the types of problem the tactic applies to.
Solution Description	A description of how the tactic achieves its objective. For example, *"This tactic uses countermeasures to distract the torpedo and then evades by turning, diving and accelerating"*.
Objective	The goal that the tactic achieves.
Context Condition	The specific context in which the tactic is applicable.
While Condition	This condition must remain true while the tactic is executing. If it becomes false at any point, the tactic is dropped.
Success Condition	The conditions under which the tactic succeeds.
Actions	Performed by the tactic.
Belief Sets	Used by the tactic.
Messages	Sent to or received from an agent/team.
Percepts	Received by the tactic.
Plans	The procedural methods used to achieve the tactic's objective.
Tactics	Its sub-tactics.

Table 2.8 Tactic properties

2.21 Team Tactic Artefact

As was discussed in Section 1.4, the pursuit of a joint goal by its members is the key property that makes a team different from other more general organisational structures. In order for the team members to effectively coordinate their joint pursuit of the team goal, they need to have a common understanding of what their approach will be. This common view of the tactical approach is represented in TDF by a **team tactic**, which is similar to an agent tactic, but team tactics use team plans instead of agent plans, the recipient of a message (or delegated goal) is always a role, and they do not receive percepts or perform actions[9].

2.22 Summary

This chapter began with a brief outline of the TDF methodology and went on to present a detailed account of its notation. The next three chapters will expand on

---

[9] The agents in the team are responsible for perception and action.

the methodological aspects of TDF and will introduce the diagrams that relate to the Requirements, Architecture, and Behaviour stages. This will be followed by a tutorial example in Chapter 6 that illustrates the various concepts introduced in earlier chapters.

Chapter 3
Requirements Stage

The Requirements stage comprises three main phases: (i) Scope Requirements (Section 3.1), (ii) Knowledge Elicitation (Section 3.2), and (iii) System Specification (Section 3.3). Scoping requirements involves determining the objectives of the modelling exercise. Knowledge elicitation leverages cases studies to determine the decision making that occurs in particular scenarios. System specification takes the elicited information and represents it in a more general, abstract diagrammatic form that can be used as the basis for designing artificial agents, such as autonomous systems, or virtual characters in simulation environments. Figure 3.1 shows the three main phases of the Requirements stage and the types of diagram used in each phase.

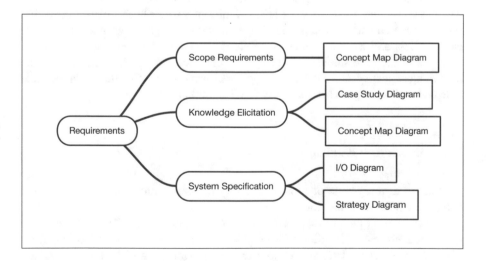

Fig. 3.1 Phases and diagrams of the Requirements stage

© Springer Nature Switzerland AG 2019
R. Evertsz et al., *Practical Modelling of Dynamic
Decision Making*, SpringerBriefs in Intelligent Systems,
https://doi.org/10.1007/978-3-319-95195-9_3

3.1 Scope Requirements

This phase focuses on the requirements of the modelling exercise, including prerequisite knowledge required by the interviewer, and the number, duration and structure of the interview sessions. It results in an initial concept map diagram that focuses on the modelling objective (for example, Figure 6.3) and comprises the following four steps:

- **Elicit Stakeholder Objective.** If relevant, interview the stakeholder to determine the purpose of the modelling exercise; for example, the stakeholder might want to answer the following question: *'How many escort ships are required to protect an aircraft carrier from submarine attack?'*.
- **Elicit Modelling Objective.** The stakeholder objective is usually not sufficiently specific for the purposes of modelling, and it needs refining to derive a concrete **modelling objective** – ideally a testable hypothesis that relates independent to dependent variables. If the modelling objective can be expressed quantitatively, determine the **MOEs**[1].
- **Study Background Material.** To ensure the interviewer has sufficient grounding in the domain to efficiently interview the domain expert.
- **Scope Interview Sessions.** The purpose is to determine how much time is available and to plan the sessions so that the stakeholder objective is adequately addressed.

The modelling objective is central to the TDF elicitation process and is used to ensure that the interview does not drift off into interesting but irrelevant detail. As you will see in the next section, knowledge elicitation is centred around one or more case studies, and a concept map is developed to show the relationships between concepts in the case studies. To ensure that the elicitation process focuses on the modelling objective, the modelling objective forms the root of the initial concept map and is linked to the MOEs, if any (see Figure 6.3).

3.2 Knowledge Elicitation

This phase develops one or more case study diagrams and fleshes out the initial concept map from the previous phase.

There are many possible approaches to knowledge elicitation; Burge [10] presents a comprehensive classification of general knowledge elicitation techniques that remains largely relevant today. A more recent taxonomy of knowledge elicitation

---

[1] An **MOE** (Measure of Effectiveness) is a quantitative impact assessment, e.g. *'Dollar value of aircraft carrier damage saved per escort ship deployed'*. In simulation applications, the values for a dependent variable are typically aggregated across multiple simulation runs to yield an **MOP** (Measure of Performance), e.g. *'average dollar value of aircraft carrier damage'*. The MOE combines the MOPs into an impact statement.

methods can be found in [55], which has a large overlap with the methods listed by Burge. The wide-ranging knowledge elicitation methods listed in these taxonomies have different strengths and weaknesses that depend upon characteristics such as interviewer skill, the ability of the expert to verbalise the knowledge, and the type of knowledge sought. We have chosen to develop an interview-based elicitation methodology for TDF, as this appears to be the most efficient method in terms of time and resources [17].

3.2.1 Specify Case Study

Experts typically find it difficult to describe tactics in the abstract. To ameliorate this difficulty, the TDF methodology encourages the use of one or more concrete case studies to provide a focus for the elicitation. The domain expert is asked to recount a case study that relates to the modelling objective. Ideally, the domain expert was the primary decision-maker in the case study; failing that, they should recount a familiar case study or generate a hypothetical one. To keep the elicitation session flowing smoothly, the case study is initially jotted down as an informal narrative (the **situation description** and **synopsis**, Table 2.1), but it is later recast in a case study diagram, as was shown in Figure 2.7. This informal narrative begins with a *situation description* that outlines the key features of the situation, for example:

> BLUE STG (Surface Task Group) is tasked with destroying the RED base. We expect RED to have dispatched a submarine to intercept. RED's best tactic is to intercept STG in the Strait, a significant choke point. If the RED submarine reaches the Strait, it will be a threat to STG. STG comprises a destroyer protected by two escort ships and a helicopter with dipping sonar.

The case study narrative continues by specifying the synopsis – a sequence of the significant events that occur in the case study, for example:

1. STG travels towards the RED base using a zigzag pattern and active sonar according to a *2-ship plus one helicopter* sonar policy.
2. An escort ship detects an incoming torpedo.
3. STG aborts the mission and returns to base.

3.2.1.1 Case Study Diagram

The next step is to flesh out the synopsis by creating a case study diagram. However, before doing so, it is prudent to verify that the case study addresses the modelling objective. This will help ensure that time is not wasted developing a case study diagram only to find that it does not help with the modelling exercise. If the case study does not adequately address the modelling objective, a new one should be selected before the case study diagram is drawn.

The nodes and artefacts in a participant row are connected together by solid **sequence** arcs that are directed from left to right. Because a case study diagram represents a chronological sequence of events, there are no loops. Dashed directed arcs cut across rows to convey messaging between participants. Once the case study diagram has been drafted, prompt the domain expert to identify the key cognitive elements underlying the decision making, for example, the goals adopted/dropped/considered, and the situation awareness requirements. These are added to the case diagram by attaching notes to the relevant artefacts and nodes, and the most important aspects to identify are:

- Goals adopted/dropped/suspended/resumed and the conditions under which this happens.
- The situation awareness needed to make a decision.
- Timing information (relative and/or absolute) that relates to the events in the case study.

In considering alternative courses of action, the domain expert may think of significant variations to the case study. Minor variations can be expressed by adding notes, but significant variations should be drawn in a new case study diagram. It is important to keep the diagrams uncluttered and easy to understand. To this end, any tool that implements the TDF methodology ideally should support the folding and unfolding of episode columns in the diagram. This is illustrated in Figure 2.7, with the `Defend Against Submarine` episode folded.

3.2.2 Populate Concept Map from Case Study

The concept map is developed in collaboration with the domain expert and acts as a shared diagrammatic representation of the domain. It does not constrain the interviewer to a particular subset of TDF's artefacts, and this notational freedom is instrumental in fostering rapid knowledge elicitation. However, TDF encourages the addition of structure to the concept map after each elicitation session by identifying relevant artefacts (see Section 3.2.3). Note that TDF concept maps do not require arc relationships to be labelled; this is intended to facilitate rapid concept map construction.

In TDF, the modelling objective forms the root of the concept map. Although the analyst is free to develop disjoint maps separately, they should ideally be linked to the modelling objective so that the analyst is reminded to consider the extent to which the elicited knowledge relates to the modelling objective. Concept maps tend to be hierarchical, with the most important concepts appearing towards the root of the hierarchy.

Development of the concept map should begin by exploring the modelling objective. Other concepts are then added, each labelled with no more than a few words; if sentences are used as labels, it implies that the concept needs to be partitioned into smaller components. Typically, a concept should only appear once in the map.

Although the layout may be clearer if the concept is represented more than once in the map, it is preferable to reduce the duplication of concepts as much as possible. Also, having a large fan of concepts emanating from a single concept can indicate that the concept needs to be partitioned into sub-concepts.

3.2.3 Classify Concept Map Nodes

The concept map developed during knowledge elicitation comprises a set of undifferentiated (generic) node types, connected by directed arcs that can optionally have an associated arc label. Once the elicitation is over, it is beneficial to classify those nodes that relate to TDF concepts, such as *goals*, *percepts* and *actions*. This facilitates the development of TDF design diagrams, brings out the tactical aspects, and also highlights concepts that are unclear or underspecified (see Figure 2.8). Goals in the concept map are identified and annotated with the events that trigger their adoption, as well as the methods used to achieve those goals, and the conditions that indicate that a goal should be dropped in favour of a more pressing one. Typically, this classification process will occur after the interview, although the domain expert may want to be involved in the exercise, which can be advantageous. The classification process results in a partial TDF design with any gaps identified and comprises the following steps:

- Identify TDF artefacts in the concept map, and label accordingly.
- If required, break high-level concepts into their component TDF artefacts.
- Identify goals and missing contextual information that specifies the conditions under which they should be adopted and/or dropped.
- Mark any further concepts that are underspecified or unclear (to be addressed at the next knowledge elicitation session).
- Identify concepts in the case study relating to goals, tactics, roles, teams, agents, key conditions, and domain-specific terminology.
- Build a partial TDF design using the identified artefacts. Typically, there will not be enough information to produce a full TDF model after the first interview, but the process of mapping out an initial design will help highlight what needs to be addressed in the next interview session.

3.2.4 Assess Results and Plan Next Elicitation Session

Once the concept map nodes have been classified, and gaps identified, the next elicitation session should be planned to target unclear aspects and missing information.

- **Purpose.** To identify missing information and prepare to fill the gaps in the next elicitation session.
- **Method/Steps.**

   - Work through any artefacts that were marked as underspecified during the
     previous stage.
   - Verify that there is sufficient information to build the relevant strategies.
   - Verify that there is enough information to build the plans and tactics required
     to achieve the goals in the strategies.
   - Identify missing artefacts, for example, percepts that are needed to discrimi-
     nate between alternative plans for achieving a goal.
   - Work through all identified gaps; draft questions for the next elicitation ses-
     sion.

- **Outputs.** Areas to focus on in the next elicitation session, in particular, any gaps
  identified.

3.3 System Specification

Having scoped the requirements, and elicited the key information about the domain to be modelled, it is time to develop a high-level specification of the decision-
making system to be built. What inputs should the system handle? What outputs
should it produce? What objectives should it be able to achieve? What high-level
strategies can it use to achieve its objectives? The inputs and outputs are represented in an I/O diagram and the objectives and strategies are specified in strategy
diagrams.

3.3.1 I/O Diagram

The I/O diagram maps inputs and outputs to the case studies that were developed
during the knowledge elicitation phase. Inputs are represented by percepts and outputs by actions. For a given case study, the percepts and actions are connected to the
appropriate external entities. The environment can be modelled as a single external
entity or more richly as a number of external entities. In general, the latter approach
is preferable because it helps with visualising the entities in the environment and
how they will interact with the system.

For example, Figure 3.2 illustrates the I/O diagram for a case study showing the
interactions of the submarine commander with four other entities on the submarine.
The four entities are represented as *external* entities because they are not modelled
as agents in the system. In this case study, the submarine commander interacts with
the 'Navigation Station' in order to move the submarine into an attacking position. The 'Navigation Station' provides updates on the submarine's depth,
location and course, and allows setting of course, speed and depth.

The 'Sonar Station' stows the towed array before the attack, as it will compromise manoeuvrability. Active sonar may be used to obtain accurate target information via the 'emit ping' action, yielding target bearing and target range.

The 'TMA Station' employs **TMA** (Target Motion Analysis) to determine target course, speed, bearing and range. TMA data is inherently error-prone, and so the 'TMA Station' has been separated from the 'Sonar Station' to make the diagram clearer.

The 'Fire Control Station' allows one to designate the target to attack, using targeting information derived from the 'Sonar Station' and 'TMA Station'. Before firing, one must select the firing tube and configure the torpedo (for example, set the target depth, torpedo floor and ceiling, and other parameters). One must also set run-to-enable, which is the distance to travel before the torpedo turns on its active sonar to home in on the target.

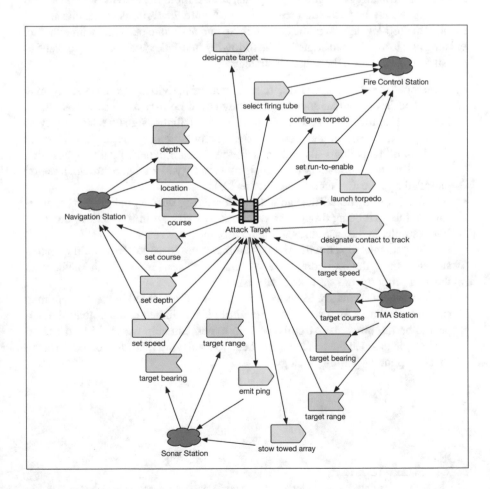

Fig. 3.2 I/O diagram

3.3.2 Strategy Diagram

Recall from Section 2.8 that a strategy is a hierarchical structure comprising goals that can be annotated by conditions that specify the life cycle of the goals, and arcs that can be annotated with the conditions under which one sub-goal should be preferred over its siblings. In addition to goals, the strategy diagram can also show any tactic that relates to one of the goals in the strategy by virtue of having that goal as its objective. This is shown in the strategy by linking the goal in question to the tactic and is illustrated in Figures 6.14 and 6.15.

As explained in Section 3.2.3, during the process of concept map classification, some nodes will likely be identified as goals and annotated by the conditions under which they might need to be dropped etc. Furthermore, the activities in case studies will typically map naturally to goals, for example the activities `Travel to conflict zone` and `Return to base` in Figure 2.6. At this stage, one should identify the top-level goals in the concept map and associate each one with either an existing or a to-be-defined strategy. The following heuristics[2] can be used to identify a goal that should form the root of a strategy:

- **Sequence of related activities.** A sequence of activities in a case study may be related in a way that suggests that they could be part of an overall strategy. For example, the HVU participant in Figure 2.7 first assigns a sonar policy to its team members, and then adopts a zigzag navigational pattern. Knowledge of the domain, or perhaps discussion with the domain expert, might suggest that these activities are sequential sub-goals of a strategy to navigate through a region patrolled by a hostile submarine (Figure 3.3).
- **Activity interruption.** This can indicate a strategy with a goal property that results in the goal being dropped in response to a change in the tactical situation. For example, in the case study in Figure 2.6, the `Travel to conflict zone` activity gets interrupted when there is an incoming torpedo. This could map to a strategy where the root goal, `Travel to conflict zone`, is annotated with the goal property `WHILE: NOT Incoming torpedo`.
- **Goal/sub-goal structure in concept map.** A goal identified in the concept map can be a candidate for a strategy, particularly if it is linked to other goals that appear to be its sub-goals. Also, the domain expert can help with the identification of strategic goals in the concept map, i.e. by identifying goals that could form the root of a general purpose strategy.

---

[2] Note that these are only *guidelines* for identifying goals and strategies to be modelled in TDF. Regardless of methodology, a certain degree of subjectivity will always be involved in any modelling exercise.

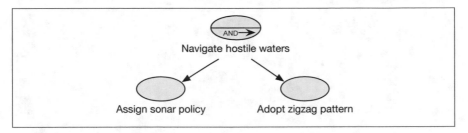

Fig. 3.3 Strategy to navigate hostile waters

Note that TDF's focus is dynamic decision making, and so it supports and encourages the capture of goal-oriented dynamics in the strategy diagram using goal properties. The goal-oriented dynamics can be teased out by seeking to answer the following questions during strategy construction:

- **What situational factors constrain the adoption of a goal?** These can be captured using the `INITIAL CONTEXT` goal property.
- **What factors cause a goal to be dropped?** Captured using the `WHILE` goal property.
- **Can the goal be serendipitously achieved due to a change in the situation?** If so, add a `SUCCEED IF` condition.

Chapter 4
Architecture Stage

In the Architecture stage (see Figure 4.1) one specifies the internal structure of the decision-making system being modelled, in particular the system roles that the agents and teams take on (in the role enactment diagram), the team structures (in the team structure diagram), and the relationships between the constituent agents, teams and other artefacts (in the architecture overview diagram).

The role enactment diagram captures the major system roles, what goals a role player needs to be able to achieve, and which agents and/or teams can enact which roles. The team structure diagram specifies the hierarchical configuration of the team and how that configuration is affected by situational factors. The architecture overview diagram maps out the agents and teams, and their interactions with one another and the environment (goals, messages, percepts and actions).

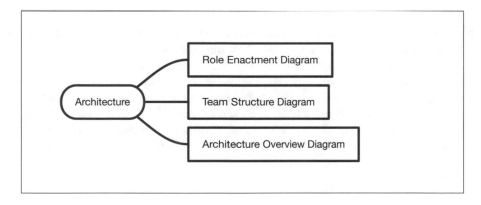

Fig. 4.1 Diagrams of the Architecture stage

R. Evertsz et al., *Practical Modelling of Dynamic Decision Making*, SpringerBriefs in Intelligent Systems, https://doi.org/10.1007/978-3-319-95195-9_4

4.1 Role Enactment Diagram

Once some strategies have been identified during the System Specification stage, one should consider who will take on the responsibility for achieving the goals that make up the strategies. To do so, one needs to identify the various role players in the system and what goals they will be able to achieve. Roles are linked to goals in the role artefact definitions, and the role enactment diagram specifies the *types* of agent or team that can *enact* each role. These are specified by linking the roles to the agents and/or teams they are enacted by.

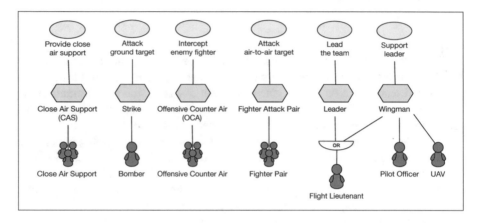

Fig. 4.2 Role enactments

Figure 4.2 shows some role enactments from an air combat scenario. Note that, in principle, a given role can be enacted by more than one entity, in this case, the `Wingman` role can be enacted by a `Flight Lieutenant`, `Pilot Officer` or `UAV` agent. Also, an entity can potentially enact more than one role; in this example, an agent of type `Flight Lieutenant` can enact *either* the `Leader` or `Wingman` roles, but never both, as shown by the *exclusive* **OR** node. If it were possible for a `Flight Lieutenant` to take both roles simultaneously in a team, then the `OR` node would be omitted and the choice would be made when the specific team instance is created, as defined by the constraints within its *team structure* definition (see Section 4.2).

4.2 Team Structure Diagram

A key feature of TDF is its support for the declarative specification of dynamic teams. Typical tactical scenarios are dynamic, i.e. the situation can change significantly and unexpectedly within a short timeframe, and so a static team structure will not suffice. Consequently, TDF not only supports the representation of the structure of a team, but allows one to specify how that structure depends on the tactical situation and the conditions under which the team should be restructured or disbanded.

Team oriented programming languages typically address this need by providing primitives for *procedurally* specifying the required team formation steps (cf. JACK Teams' team formation plans [23]). However, a team formation plan is usually very difficult for domain experts to understand because it forces them to *mentally simulate* the team formation steps in order to derive the resulting team structure for a given situation. To address this difficulty, TDF employs a *declarative* representation for team structures, which includes a means of specifying how the team structure depends on situational factors. A situational factor is expressed as an arc condition (using the `IF` keyword) between a role and the role filler, with an optional `WHILE` condition to express when the role requirements change[1].

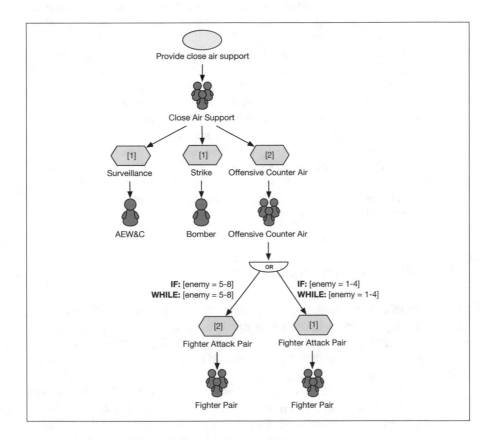

Fig. 4.3 Team structures

---

[1] A change in role requirements may lead to the enacting sub-team(s) being disbanded, or the enacting agent(s) being dropped from the team.

Figure 4.3 shows the team structure for a `Close Air Support` team and its `Offensive Counter Air` sub-team. The `Close Air Support` team comes together to handle the `Provide close air support` goal, and the team exists for as long as the goal remains active. When the decision-making system adopts this goal, it tries to find a way of achieving it, and one of those ways might require the formation of a team that can handle the goal[2].

Once the team is formed, it then adopts the goal and tries to achieve it using the **team tactics** and **team plans** at its disposal. If the goal is achieved or is dropped for some reason, the team gets disbanded. The `Close Air Support` team is made up of three roles: `Surveillance`, `Strike` and `Offensive Counter Air`; the latter requires two role fillers, as shown by the `[2]` label. These roles are respectively enacted by an `AEW&C` agent, a `Bomber` agent and two `Offensive Counter Air` sub-teams. In this context, the `Offensive Counter Air` team is a sub-team, and so, unlike the `Close Air Support` team, it does not require a team formation goal (it simply gets formed as a side effect of the formation of the `Close Air Support` team).

The `Offensive Counter Air` team will comprise two `Fighter Attack Pair` role fillers if there are between five and eight enemy fighters; otherwise, if there are between one and four enemy fighters, then only one role filler is required. The `WHILE` condition denotes that the role fillers are part of the sub-team for as long as the `WHILE` condition remains true. So, in this example, if there were five enemy fighters, two `Fighter Pair` sub-teams would be formed. If at any time the number of enemy fighters dropped below five, then only one `Fighter Pair` would be needed, and so one of the sub-teams would get disbanded. This could happen due to enemy losses during air combat or because the initial estimate of the number of incoming fighters was wrong, and the true value only became apparent once they were within visual range.

4.3 Architecture Overview Diagram

The architecture overview diagram is perhaps the most informative diagram of all because it brings together, into a single diagram, an overall view of the types of entity (agents and teams) in the system being modelled, the types of messaging between them, goal requests that are sent between entities, and how the agents relate to incoming percepts and outgoing actions.

In Prometheus [49], the methodology upon which TDF is based, messaging can also be structured into interaction protocols that specify the sequential ordering of the messages, using Agent UML [5]. This works well for typical software engineering applications, where static interaction protocols are quite common, but we have found them to be less useful for modelling dynamic decision making. Although military domains can involve predefined protocols between participants, such protocols

---

[2] In other cases, the goal might simply be handled by an agent armed with an appropriate tactic or plan, without having to form a team to tackle the goal.

are, by definition, fairly fixed and so tend to lie outside the dynamic aspects of the system. Transparent and effective capture of *dynamic* protocols is an open area of research, which we are working towards incorporating in a future revision of TDF.

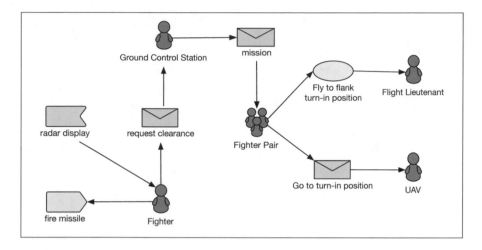

Fig. 4.4 Architecture overview

Figure 4.4, although showing only a small part of an architecture overview diagram, illustrates all of the types of relationship that can be represented therein. Working through the diagram top-down and left-to-right, the `Ground Control Station` agent can receive a `request clearance` message from a `Fighter` agent, which takes in `radar display` percepts and can execute a `fire missile` action; the `Fighter Pair` team can receive a `mission` message from the `Ground Control Station` agent, and can send a `Go to turn-in position` message to a `Flight Lieutenant` agent; it can also delegate the `Fly to flank turn-in position` goal to a `UAV` agent.

Chapter 5
Behaviour Stage

In the Behaviour stage one focuses on the behavioural aspects of the system entities and the details of their internal structure. Agent diagrams and team diagrams show the internal behavioural components of agents and teams. At the macro level, the behaviour of agents and teams is specified in terms of tactics, and these are shown in their respective tactic and team tactic diagrams. The detailed behavioural steps that an agent or team follows are captured respectively in plan diagrams and team plan diagrams. Figure 5.1 summarises these six types of Behaviour stage diagram.

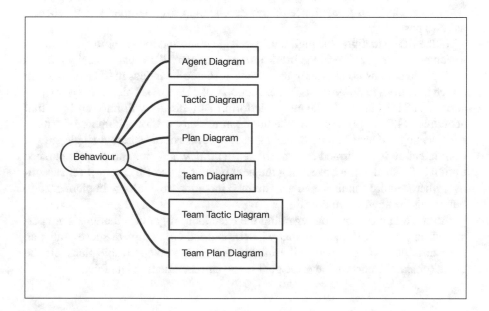

Fig. 5.1 Diagrams of the Behaviour stage

© Springer Nature Switzerland AG 2019
R. Evertsz et al., *Practical Modelling of Dynamic Decision Making*, SpringerBriefs in Intelligent Systems,
https://doi.org/10.1007/978-3-319-95195-9_5

Because TDF is a design methodology rather than a programming language, it is agnostic with regard to the specifics of agent execution. Nevertheless, TDF is founded on the BDI paradigm, and so it comes with some general assumptions about how its models will map to a BDI execution engine. At an abstract level, the agent execution cycle follows a typical BDI loop as follows:

1. Observe new inputs and decide on which events to respond to. The events could be external (percepts received from the environment) or internal (goals/messages generated by the agents or teams).
2. Determine the options the agent can pursue in response to the events. In practice, these options are plans that handle the events. A plan is chosen as an option if the plan is *relevant* and *applicable*.
3. Deliberate over these options and select which ones to execute next. This deliberation depends on the agent platform and may include mechanisms for conflict resolution [62, 60], scheduling [61, 71, 72], aborting and suspending goals [59, 34] and so on.
4. Drop completed plans as well as goals and plans deemed to be impossible.

In Step 2, there could be cases where more than one plan is applicable (i.e. their context conditions are all satisfied). TDF does not specify which plan should be selected; this would depend on how the target implementation platform chooses between applicable plans. For example, an agent platform such as JACK [23] uses factors such as **plan precedence** to choose one plan over another. Other schemes are also possible.

In the BDI paradigm, the high-level strategic goals and the specific plans chosen to achieve those goals form the **intentions** of the agent. An agent typically pursues multiple intentions concurrently and, when resource-bounded, may have to select which intentions to progress at each execution cycle (Step 3). This choice is not prescribed in TDF but left to the agent platform that typically will have an **intention scheduler** [42]. In principle, a scheduler can use any one of a number of criteria for selecting the next intention, including (i) **round robin**, where a single step of each intention is performed in turn; (ii) **sequential**, where one intention is pursued until it is finished, before executing the next intention; and (iii) **meta-level reasoning**, where a more sophisticated user-defined intention scheduler is implemented to better suit the application domain.

An important assumption regarding the execution of goals is that goals are persistent. In practice, this means that when a plan fails to complete successfully, an alternative plan will be selected if available. If there are no applicable plans left, the agent (or team) cannot achieve the goal in the current situation and fails.

5.1 Agent Diagram

The agent diagram (Figure 5.2) shows the properties of the agent artefact. The agent's plans determine which actions, percepts, goals, and messages it has, and so the diagram centres around the agent's plans, with the other artefacts shown connected to the plans they belong to. The diagram also shows the agent's tactics; for each tactic, it shows the tactic's objective (a goal of the agent).

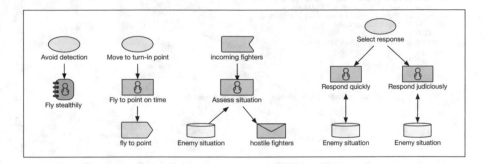

Fig. 5.2 Part of the agent diagram for a combat pilot

5.2 Tactic Diagram

As explained in Section 2.20, the tactic artefact is an encapsulation mechanism that allows functionality to be aggregated for the purposes of reuse as well as the explanation of agent behaviour. The tactic diagram allows one to assign relevant artefacts to the tactic (see Table 2.8). Figure 5.3 shows part of the tactic diagram for a tactic that can search for a target and, for the purposes of illustration, includes all of the artefacts a tactic can have. Recall that a tactic also incorporates aspects of design patterns, such as a problem description.

The objective of the tactic is 'Search for target'. It can make use of two sub-tactics, 'Expanding Square Search' and 'Creeping Line Search', each of which is a different way of handling the goal 'Adopt search strategy'. The goal 'Record position history' is handled by the plan 'record search leg parameters'.

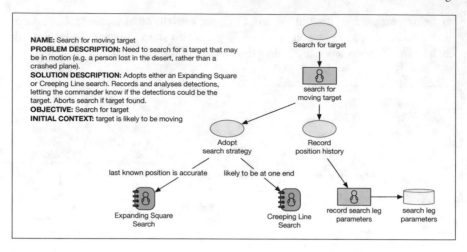

Fig. 5.3 Tactic Diagram

5.3 Plan Diagram

In TDF, plans are specified diagrammatically; plan diagrams were illustrated in Figure 2.10. Each plan has a unique name and is defined, edited and viewed in its own diagram. Once defined, plans can be assigned to tactics and agents.

5.4 Team Diagram

A team diagram is similar in content to an agent diagram, but it substitutes team plans and tactics for agent ones, and excludes percepts and actions. An example team diagram is shown in Figure 6.28. Note that it does not show the team's members because these are defined in the team structure diagram (Figure 4.3) and can vary depending on the situation. The team diagram only shows the *fixed* properties of the team artefact.

5.5 Team Tactic Diagram

Like agent tactics, the team tactic artefact (Section 2.21) is an encapsulation mechanism that enables functionality to be aggregated, reused, and employed to explain team behaviour. Because a team cannot directly perceive or act (it can only do so through its member agents), the team tactic diagram does not include percepts or actions.

Figure 5.5 shows part of a team tactic diagram that relates to a pincer attack
(Figure 5.4). The team tactic's objective is 'Attack fighters'. It employs the
team tactic 'Formation flying', which has the objective 'Fly in formation'.
It has two plans that handle the goal 'Implement pincer'; one that handles the
goal when the target is operationally critical and the other that handles the goal
when the target is optional. The two plans can read from and write to the belief
set 'Enemy situation', as denoted by the bidirectional arcs, and can delegate the
goal 'Fly to flank turn-in position' as well as send the message 'Go to
turn-in position'.

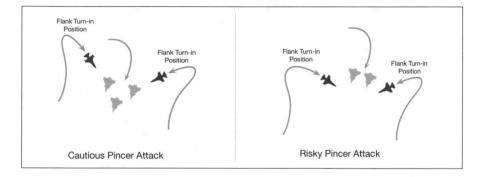

Fig. 5.4 Pincer attack

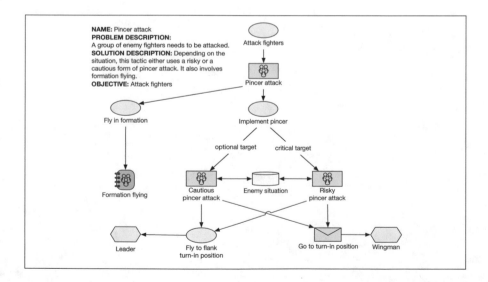

Fig. 5.5 Team Tactic Diagram

5.6 Team Plan Diagram

As explained in Section 2.18, team plans are quite similar to agent plans, but they have particular restrictions and extensions that make them more suitable for modelling the decision making of teams. They are defined, edited and viewed in team plan diagrams.

Chapter 6
TDF Tutorial Example

This chapter presents a tutorial on how to develop a model in TDF. Different domains have different characteristics, and this will affect the degree to which various TDF features are needed. For example, a domain such as air combat is highly team-oriented and tactically fast moving. In contrast, undersea warfare moves at a glacial pace and is not as rich in team-oriented tactics. We have selected a realistic example, based on Akbori's master's thesis [4], and have adapted it to highlight key aspects of TDF. Akbori used various scenarios to investigate the application of agent-based simulation to the problem of how best to protect an HVU (High Value Unit) from submarine attack, using a number of escort ships. In his case study, the submarine works alone, but the ships work together as a team, and this latter aspect offers a good opportunity to demonstrate team modelling in TDF.

In this chapter, we will work through the Akbori case study in order to add flesh to the bones of the methodology described in earlier chapters. Although it is not possible to present a complete design within the confines of this book, we shall endeavour to provide thorough examples of the important aspects of TDF. Hopefully, this will give you a solid grasp of the TDF methodology and how to develop clear and effective models of dynamic decision making.

To reduce the need to refer back to previous chapters, the overall TDF process is summarised in Figure 6.1, and this chapter reflects that structure. Before working through the case study, we present some background information on undersea warfare in order to provide the reader with a better understanding of the domain.

© Springer Nature Switzerland AG 2019
R. Evertsz et al., *Practical Modelling of Dynamic
Decision Making*, SpringerBriefs in Intelligent Systems,
https://doi.org/10.1007/978-3-319-95195-9_6

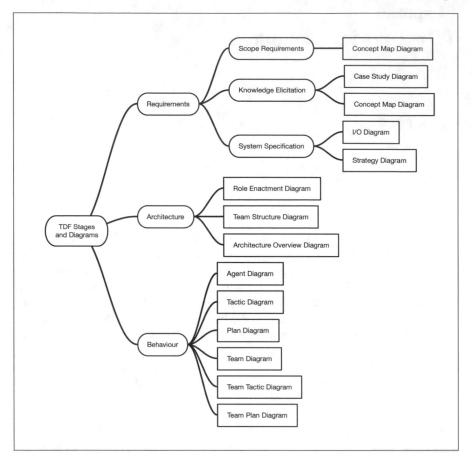

Fig. 6.1 TDF process and diagrams

6.1 The Undersea Warfare Domain

In order to fully appreciate some of the subtleties of the case study presented here, it is important to understand the key properties of the undersea warfare domain. The tactical approaches only make sense if one appreciates how uncertain the situation is from the submarine's perspective, and how the escort ships and HVU work together to obfuscate the submarine's situation awareness.

The undersea warfare domain is distinguished by the paucity of information about the current situation. A submarine commander's knowledge of the tactical situation is time-consuming to acquire, very limited and is sometimes highly uncertain. This severely impedes the decision-making process – much of a commander's tactical repertoire is concerned with building situation awareness whilst not being detected. This places constraints on the use of sensors and the types of manoeuvre available.

On a submarine, sonar is the most important class of sensor. Due to the requirement to remain concealed until it is time to attack, a submarine commander mostly relies on passive[1] rather than active sonar because the latter could reveal his presence. Unfortunately, the undersea environment is noisy and heterogeneous (sound may not travel in straight lines); underwater sound bends, is reflected, can get trapped in a thermal layer, can be split by a thermal boundary and can reverberate. Furthermore, sea state can create noise that degrades the signal, as can the submarine's motion through the water – faster speed produces noise that blurs the sonar picture while also exposing the submarine to detection. The speed of sound through water varies with depth, temperature and salinity. This variation in transmission speed causes refraction; thus, when a contact is picked up on passive sonar, one cannot be sure of its bearing. TMA (Target Motion Analysis) [15] is used to abductively infer the contact's bearing, range and course, but it is an inexact science at the best of times. At the very least, a given contact can have two interpretations: the contact is **[near, quiet, slow]** or **[far, loud, fast]**. Naturally, there is a continuous range of combinations in between these two boundaries – all are valid interpretations. TMA is used to narrow the set of possibilities. The implication for undersea warfare tactics is that one can rarely be confident in one's appraisal of the situation if it is based on passive sonar, but because of the need for stealth, many tactics are designed to covertly increase situation awareness using passive sonar alone.

Unfortunately for the submarine, active sonar is more likely to alert the adversary to its presence. Active sonar is much more accurate than passive sonar, but it betrays one's bearing, range and even identity (submarine type); thus, its use is typically limited to the final stages of an attack. Using the periscope constrains the submarine to being within about 20 metres of the surface, and limits speed to about 5 knots, but gives an instant target solution. However, its wake can alert surface ships and aircraft to the submarine's precise position. Similarly, ESM (Electronic Surveillance Measures) involve raising an antenna above the water's surface to gain target range and bearing information; however, the antenna also produces a revealing wake. Building a situation picture comes with assumptions about the effectiveness of one's own sensors in the current environment, as well as assumptions about the adversary's sensor capabilities and modus operandi. Sensors can be uni/omnidirectional, narrow/broadband, provide bearing and/or range, etc. Undersea warfare tactics take these limitations and capabilities into account, for example, by having the submarine adopt a zigzag pattern to overcome the fact that its towed sonar array is blind to targets ahead.

In contrast to other military domains, published studies of human decision making in undersea warfare are few and far between because tactics are kept secret. Studies of decision making in submarine-related tasks have focused on the biases and limitations of the human cognitive system. The only accessible study of submarine commander decision making is Project Nemo [33]. Analysis revealed a shal-

---

[1] Passive sonar sensors detect incoming sound energy. Active sonar generates a sound pulse and provides target bearing and range information when reflections are detected.

low, adaptive goal structure that is also characteristic of schema instantiation [66]. In other words, the commanders had built up a repertoire of situation descriptions coupled with tactics that worked well in those situations, rather than performing a search through a larger problem space. This approach to decision making is a trademark of expert problem solving [39] and is well represented by the BDI paradigm which tends to focus on pre-compiled recipes for problem solving, rather than deriving the solution from first principles.

6.2 The Akbori Case Study

In his thesis, Akbori explores a number of variants of an anti-submarine warfare scenario in which a group of escort ships are tasked with protecting an HVU from submarine attack. We have selected one variant for exploration in this chapter; an example in which the HVU is protected by four escort ships (see Figure 6.2). Each escort is responsible for patrolling its assigned sector, which has a constant offset from the HVU; in other words, the HVU is always in the centre of the ring formed by the four sectors. The HVU can either travel directly to its destination or, in times of high risk, adopt an apparently random zigzag motion while converging on the destination, albeit more gradually. This makes it more difficult for an adversary to manoeuvre into an attack position. The escorts also zigzag within their own sectors and alternate turning their active sonar on and off, according to an overall sonar policy that is designed to confuse the submarine, as it cannot tell for sure how many ships there are and where they are going. The inner edge of each escort's sector is some 5 km from the HVU, and the outer edge is about 10 km away. Thus, it is not easy for the submarine to tell where the HVU is; it must somehow build up a situation picture, using passive sonar and the pings being sent out by the escorts, and then find a gap in the escort screen so that it can sneak through to attack the HVU. To appreciate how difficult this is, consider the fact that the escort sending out sonar pings might be travelling perpendicular to the direction of the HVU for quite some time before ceasing its pings, at which point another escort starts sending out pings while travelling in the opposite direction. Somehow, the submarine commander has to figure out how many escorts there are, where the HVU is likely to be going, and where a gap in the screen might occur, all while avoiding detection and subsequent attack. Undersea warfare is not for the faint of heart.

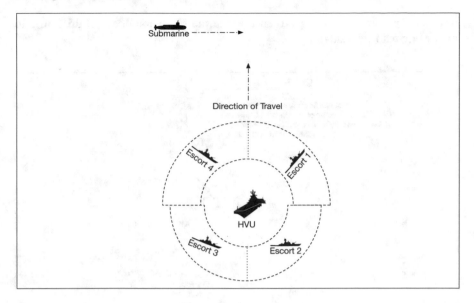

Fig. 6.2 Spatial layout of entities in vignette

6.3 Requirements Stage

Recall from Section 3.1 that the first phase of the Requirements stage is to scope the requirements. This is followed by the Knowledge Elicitation and System Specification phases.

6.3.1 Scope Requirements

The key activity of the Scope Requirements phase is to elicit the stakeholder objective and refine it into the modelling objective for the exercise. This phase results in a single diagrammatic output: the initial concept map, with the modelling objective at its root. Although we do not have an actual stakeholder to interview, we can use Akbori's stated research objective as our stakeholder objective:

> The purpose of the simulation model is to strategize Anti-Submarine Warfare (ASW) operations in order to protect a High Value Unit (HVU). [4, p. 19]

The stakeholder objective describes the overall question that needs to be answered by the modelling exercise. Often, this is not sufficiently specific for the purposes of modelling and needs refinement to derive the modelling objective – ideally, though not necessarily, a testable hypothesis. Typically, the modelling objective is the very first thing to be encoded in the concept map (Figure 6.3), and it should be positioned as the root throughout the modelling exercise so that it is kept in mind at all times. This helps to ensure that the elicitation does not drift off into interesting but irrelevant detail.

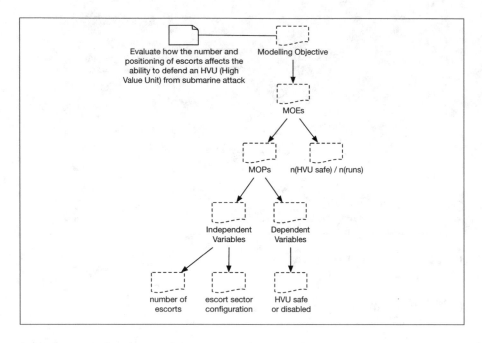

Fig. 6.3 Modelling objective

Once the modelling objective has been specified, one should familiarise oneself with the domain by studying relevant background material. The knowledge elicitation sessions should then be planned so that the modelling objective is adequately addressed.

6.3.2 Knowledge Elicitation

The Knowledge Elicitation phase of the Requirements stage results in two types of diagram: the case study and concept map diagrams.

6.3.2.1 Case Study

During the first knowledge elicitation session, the domain expert should describe a case study that addresses the modelling objective; the case study is initially written down as an informal narrative comprising a **situation description** and **synopsis**.

- **Situation description.** STG (Surface Task Group) is travelling to a conflict zone. STG comprises an HVU (such as an aircraft carrier) protected by four escort ships. An enemy submarine is likely to be encountered en route, and so STG adopts a defensive navigational configuration.

- **Synopsis.**
  1. STG travels towards conflict zone, using zigzag pattern.
  2. Escorts use active sonar according to a 4-ship sonar policy.
  3. Submarine detects pings (submarine is further than 10 km from nearest ship, and so is outside detection range).
  4. Submarine maintains a record of the sonar detections.
  5. Submarine projects STG course.
  6. Submarine calculates rendezvous point (10 km from HVU's projected position) and computes required course and speed (minimising noise and leaving at least 40% battery in reserve, to allow escape).
  7. Submarine navigates to rendezvous point.
  8. Submarine notes that STG has changed direction (due to zigzag pattern); submarine recalculates rendezvous point.
  9. At rendezvous point; HVU is estimated to be further than 10 km from submarine, but is moving towards submarine; submarine waits.
  10. Front two escorts detect submarine and move at maximum speed to intercept.
  11. Rear two escorts move at maximum speed to place themselves between HVU and submarine.
  12. HVU moves at maximum speed away from submarine.
  13. Submarine notices rapid approach of front two escorts, aborts mission, and escapes.

Fig. 6.4 Initial case study framework

Next, a case study diagram is created, in order to build a clearer picture of what the participants are doing, how they interact over time, and the potential for minor variation. However, before doing so, one should have the domain expert confirm that the case study, as specified in the synopsis, will address the modelling objective.

In general, the best approach to creating a case study diagram is to first create the participant rows and, where appropriate, group them into teams, as this will provide a framework for specifying the various threads of activity that occur during the case study. In this case study, six participants are mentioned: the HVU, four escorts, and a submarine. STG is a good candidate for modelling as a team because its members have the common goal of reaching the conflict zone. The escorts form a sub-team with the common goal of protecting the HVU and, in this case study, are further subdivided into an intercepting and a shielding sub-team because the front two escorts intercept the submarine while the rear two move to place themselves between the HVU and the submarine. This structure is shown in Figure 6.4.

The events in the synopsis are then added to the appropriate rows of the case study diagram and connected together with sequence arcs to represent the chronological ordering of the events within each row. Where needed, message and synchronisation arcs are added across rows. In Figure 6.5, a message arc is shown between the 'Assign sonar policy' and 'Adopt sonar policy' activities, and a synchronisation arc can be seen between the 'sonar pings' percept and the 'Navigate leg' activity.

Typically, once the case diagram has been drafted using the synopsis, the domain expert is asked to critique the diagram and fill out missing detail (as described in

Section 3.2.1.1). This tends to bring up aspects that were not mentioned in the case study synopsis, and it is an important part of the knowledge elicitation activity. As the details get fleshed out, new artefacts will be inserted between those from the synopsis. Because dynamic decision making tends to be very context dependent, with the help of the domain expert, one should focus on teasing out the contextually important variables; for example, the factors that would cause a participant to choose a particular course of action or drop the current course of action. Such information will be tremendously useful later on when one needs to specify the strategies, tactics and plans. These contextual aspects are added as notes, along with information about what goals the participants are trying to achieve and their situation awareness[2]; e.g. 'Main goal' and 'store course, speed, range and bearing of targets' in Figure 6.5.

Episodes should be added to demarcate meaningful phases and make the diagram more readable. The completed case study diagram is shown divided into three episodes (Figures 6.5 to 6.7) and, for illustrative purposes, some artefacts are annotated to show their relationship to the synopsis.

Figure 6.5 shows the detail of the 'Navigate First Leg' episode[3]. The HVU assigns a sonar policy to the escorts, initiates navigation by messaging the escorts with its intended velocity, and begins navigating the first leg. The escorts begin patrolling their respective sectors when they receive the 'velocity' message. At some point while the HVU is navigating the first leg, the submarine picks up sonar pings (emanating from the active sonar of the escorts); these sonar pings co-occur in time with the HVU's 'Navigate leg' activity, as shown by the dashed synchronisation arc between the two artefacts. The submarine accumulates sonar data, calculates a rendezvous point (from which to attack the HVU), and navigates to that destination.

The 'Direction Changed' episode is shown in Figure 6.6. When the HVU changes direction, it messages the 'Intercepting Escorts' and 'Shielding Escorts' teams so that their members can shadow the HVU's new direction and speed; to do so, each escort constantly updates its *virtual* sector so that it remains in the same position relative to the HVU. At some point, while the HVU is navigating the next zigzag leg, the submarine hypothesises that the HVU has changed direction and so updates the planned rendezvous point, from which to attack the HVU, and concurrently sets its target depth, heading and speed.

In the 'Defend Against Submarine' episode (Figure 6.7), one of the 'Intercepting Escorts' detects the submarine and informs the other members of STG.

---

[2] The reader may be wondering why we don't just use the artefact icons for goals, beliefs, etc., rather than annotating with note nodes. Doing so would make case study diagrams look very similar to plan diagrams, which would be fine but for the fact that the semantics are very different. This could lead to confusion when creating and interpreting a case study diagram because, unlike a plan diagram, it does *not* describe a process. Rather, a case study diagram denotes sequences of events and activities, and the interactions between participants in a specific instance.

[3] Due to space limitations, activities have been used in preference to sequences of percepts and actions. For example, the 'Patrol Sectors' activity would involve a sequence of direction and speed change actions, but including all of them would make the diagram too large to fit within the confines of the page.

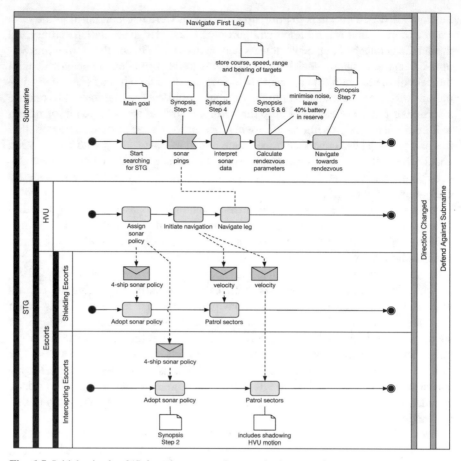

Fig. 6.5 Initial episode of 'Submarine escapes interception' case study

The HVU sets its bearing 180° away from the submarine and flees at maximum speed. Meanwhile, the 'Shielding Escorts' move between the HVU and the submarine, and the 'Intercepting Escorts' proceed to intercept the submarine.

The submarine estimates that the HVU is further than 10 km away but moving towards the rendezvous, and so it waits until the HVU is within a 10 km radius. However, before the HVU is within range, the submarine detects that the escort ships are on an intercept course, and consequently it gives up waiting and escapes.

6.3.2.2 Concept Map

As explained in Section 3.2.2, the concept map begins with the modelling objective (Figure 6.3), which is then extended with concepts that arise from the specification of the case study. To foster rapid construction, the concept map is initially created

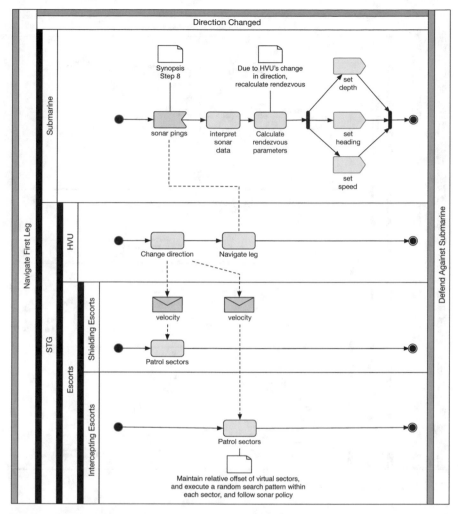

Fig. 6.6 Middle episode – HVU changes direction

using generic nodes, but the nodes are later classified in terms of TDF artefacts. In most domains, the concept map can become quite large, and so this example (Figure 6.8) only shows a very small part of the concept map that could be developed and focuses on the submarine's perspective. Concept maps are useful during knowledge elicitation because they act as a shared frame of reference for discussing the domain with the expert. They are very flexible and can be used to visually represent any type of association between the elements of the domain. This example is by no means prescriptive; concept maps can be used to sketch out all sorts of useful relationships and could, for example, be used to elicit an initial domain ontology, showing the different types of sonar and their relative capabilities.

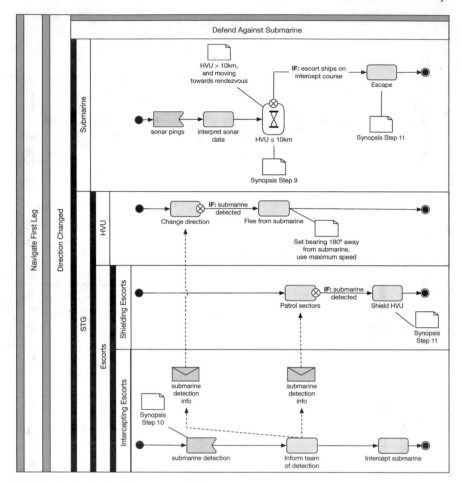

Fig. 6.7 Final episode – interception of submarine

Figure 6.8 shows some of the information required by the submarine (1). The tactic (2) involves five goals that roughly correspond to the military concept of a **kill chain**; note that the fifth goal, 'Destroy', should be dropped if it is unsafe to attack (3). The heuristics for identifying the enemy's search pattern have not been specified, and so a note is added to clarify this at the next elicitation session (4). In order to classify the target, 'broadband sonar footprint' percepts must be analysed (5). Torpedo firing is an action (6), and the estimation of the target's position is a sequence of steps (7) executed as part of the 'Get into firing position' goal.

This concludes the knowledge elicitation phase and it is now time to begin specifying the design of the model.

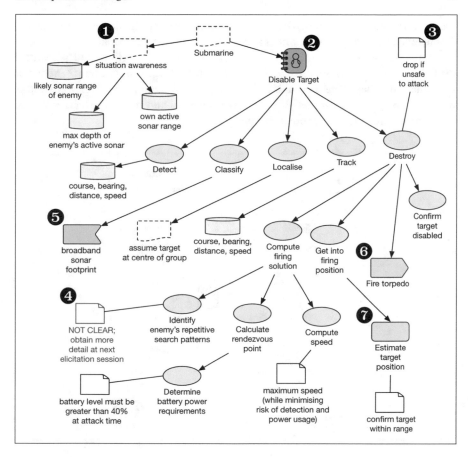

Fig. 6.8 Part of the concept map that relates to the submarine

6.3.3 System Specification

Section 3.3 outlined the System Specification phase, and introduced the two types of diagram developed, namely, the I/O and strategy diagrams.

6.3.3.1 I/O Diagram

The I/O diagram is centred around the case study in question, and it shows the incoming percepts from various environmental sources and the actions performed on the environment. If there were sufficient room to include all of the percepts and actions in the 'Submarine escapes interception' case study, the I/O diagram that was presented in Figure 3.2 would largely fit the submarine-related input/output of the current tutorial example. Recall that, in order to fit the case study diagram onto the page, the number of percepts and actions in the case study diagram was limited; this is reflected in Figure 6.9, which is consequently relatively small.

The 'set depth', 'set heading' and 'set speed' actions are performed on the 'Submarine Navigation Station', and the 'fire torpedo' action is performed on the 'Submarine Fire Control Station'. The 'Escort Sonar Station' emits 'submarine detection' percepts, and the 'Submarine Sonar Station' emits 'sonar ping' percepts.

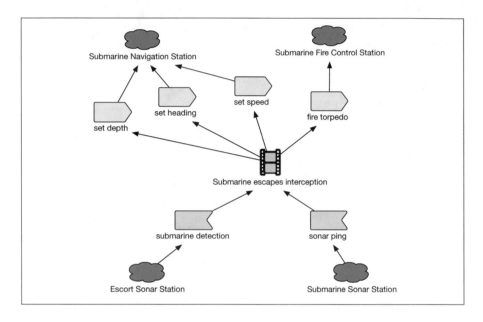

Fig. 6.9 I/O Diagram for case study

6.3.3.2 Strategy Diagram

Section 3.3.2 outlined heuristics for developing strategies from case studies and concept maps. We now present seven strategy diagrams, four for the submarine (Figures 6.10 to 6.13) and three for STG (Figures 6.14 to 6.16). Typically, strategies should only be a few levels deep, but it all depends on the particular example; it is sometimes better to have a deeper strategy tree if that clarifies the overall intent. Strategies are intended to be a high-level abstraction that can encompass a number of case studies, and they more precisely model the goal-based decision making that was only loosely sketched out in the concept map. As a given strategy is developed, one may want to go back to the concept map and rename goals to match, so that there is traceability between the concept map and the strategies[4].

Neutralise enemy (Figure 6.10): Based on the `Disable Target` tactic in the concept map (Figure 6.8), this strategy spans the three episodes of the case study (Figures 6.5 to 6.7). This strategy tries to achieve the `Detect`, `Classify` and `Disable target` goals in left-to-right order. To `Detect`, it tries to achieve the `Choose search strategy` and `Execute search strategy` goals. A strategy for achieving the goal `Disable target` is shown in Figure 6.11.

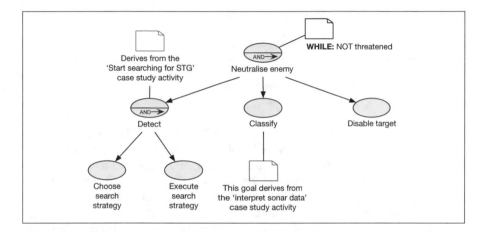

Fig. 6.10 Submarine's strategy for neutralising the enemy

Disable target (Figure 6.11): This strategy relates to the `Disable target` goal in the `Neutralise enemy` strategy. Here, the submarine concurrently performs `Track target` while trying to achieve the `Locate and neutralise` goal. Note that the `Track target` goal should succeed once `target disabled` is true. Without the inclusion of that goal property, it might not be clear that the submarine does not `Track target` forever[5].

---

[4] For example, in developing the strategy in Figure 6.10, the concept map goal `Destroy` was renamed to `Disable target`. In such cases, the TDF Tool will automatically rename the `Destroy` goal in the concept map to `Disable target` so that they match.

[5] Note that the `SUCCEED IF` goal property serves as documentation. It could be left out, under the assumption that the plan that achieves the `Track target` goal succeeds as soon as `target disabled` becomes true, but then this would not be clear to anyone reading the strategy; they would have to look at the plan definitions to determine that this would be the case.

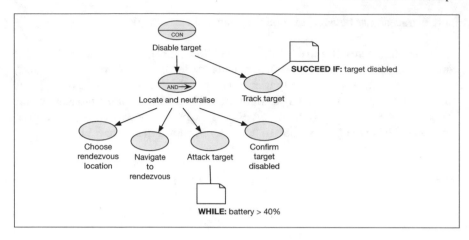

Fig. 6.11 Submarine's strategy for disabling the target

Evade torpedo (Figure 6.12): The TDF methodology has been designed to be
flexible and responsive to the needs of each modelling situation; thus, one is not
forced to create a case study or concept map for each aspect to be modelled. In
some cases, one might be augmenting the models using a Standard Operating Procedures manual. This is reflected in the inclusion here of the submarine's strategy
for evading an incoming torpedo. This event is not part of the case study or concept
map, but a strategy could be created to handle it, perhaps because the domain expert
suggested that it is relevant and is something that any submarine commander should
be able to handle. In this strategy, the submarine evades the incoming torpedo by
concurrently launching two types of countermeasure and then manoeuvring away
from them.

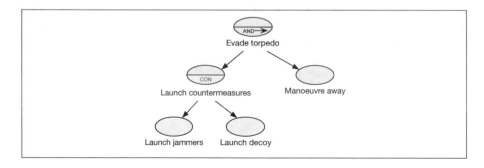

Fig. 6.12 Submarine's strategy for evading torpedo

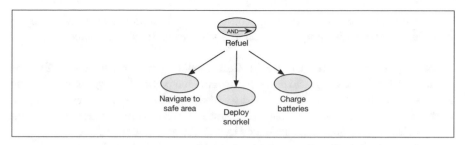

Fig. 6.13 Diesel-electric submarine's strategy for charging batteries

Refuel (Figure 6.13): In the Akbori case study, upon which this tutorial example is based, the submarine aborts its attack if it doesn't have more than 40% remaining battery charge; this is to ensure that it has sufficient endurance to escape the inevitable counterattack. The strategy diagram shows that, to refuel, the submarine first navigates to a safe area, deploys its snorkel (so it can run its diesel engine) and then charges its batteries.

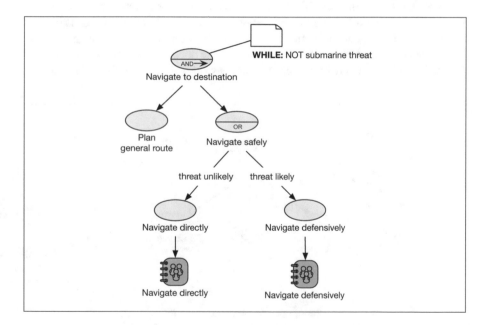

Fig. 6.14 STG navigation strategy

Navigate to destination (Figure 6.14): Here we present the strategy for navigating to a destination. It navigates directly when there is no likely threat, using the 'Navigate directly' tactic. When a threat is likely, it employs the

'Navigate defensively' tactic to confuse the submarine. The 'Navigate to destination' goal is dropped as soon as a submarine threat is detected.

Zigzag along route (Figure 6.15): Zigzag navigation involves determining the parameters for the next leg, and then concurrently navigating the leg while patrolling sectors. Once the 'Travel for leg duration' goal has been achieved, it again adopts the 'Zigzag along route' goal, and so repeatedly navigates a new leg until it reaches the destination (as indicated by the 'SUCCEED IF: at destination' goal property). This strategy pertains to the overall STG team, and so it includes the goals of the various team members in the one diagram. During the System Specification phase, we are not interested in specifying how the goals are distributed across the team members; this is modelled in the Architecture stage, in terms of the various roles in the team and what goals those roles can support. Looking at the case study diagram, we can infer that the escorts will adopt the sector patrol goal, whereas the HVU is responsible for directing the overall zigzag pattern of STG. However, one is not restricted to modelling the case study in terms of the concrete entities therein. For example, some of the HVU's responsibilities could be assigned to the STG team, which is a conceptual abstraction that has no concrete counterpart in the real world. We will model it in this way so that the zigzag pattern is handled at the *team* level, rather than at the HVU *agent* level (see the role enactment diagram, Figure 6.17). This team-level approach provides a more abstract conceptualisation of the decision making, and it particularly suits applications where the team as a whole is represented as a software module that makes the decisions for the whole team, for example, a team of Unmanned Aerial Systems working as a coordinated group.

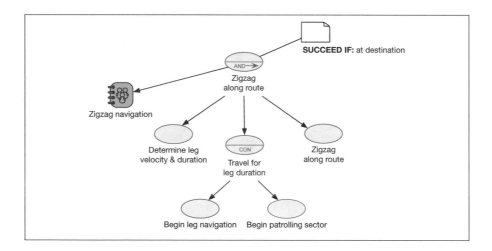

Fig. 6.15 STG zigzag navigation strategy

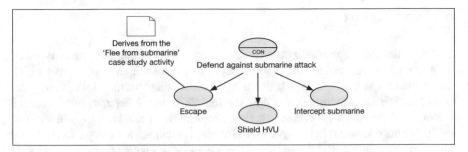

Fig. 6.16 STG strategy to defend against submarine attack

Defend against submarine attack (Figure 6.16): This strategy involves concurrently escaping, shielding the HVU and intercepting the submarine. The diagram does not specify which roles in the STG team will take on which goals. However, from the case study diagram, we can infer that the HVU will escape while one group of escorts shields it and the other escort team intercepts the submarine. Although there is no formal relationship between these goals and the activities in the case study diagram, the strategy was created with the case study in mind, and so it is annotated to highlight one of the links between the two.

6.3.4 Summary

This section outlined an example of the Requirements stage of the TDF methodology and its three phases. The Scope Requirements phase resulted in a concept map describing the modelling objective. The Knowledge Elicitation phase resulted in a case study diagram and related concept map. In the System Specification phase, an I/O diagram was created to summarise the system's inputs and outputs, and high-level strategies were created to drive the forthcoming modelling in the Architecture stage and the Behaviour stage.

Due to space limitations, in the remainder of this tutorial we will focus on the STG team rather than the submarine, and even then there is only room to present a subset of the full design of the STG team and its members.

6.4 Architecture Stage

The Architecture stage was introduced in Chapter 4 and comprises three types of diagram: (i) the role enactment diagram that lists the roles in the system and which agents/teams can enact them; (ii) the team structure diagram that declaratively defines how the team is formed and how the team's structure depends on situational factors; and (iii) the architecture overview diagram that provides a high-level view of the agents and teams in the system, the percepts handled, actions performed, and any messaging and goal delegation that can occur between agents, teams and roles.

6.4.1 Role Enactment Diagram

The goals specified in the strategy diagrams will be pursued by agents and/or teams in the system. Individual goals can be assigned to agents or teams later on, during the Behaviour stage, but the relationship between agents/teams and goals can often be made clearer by grouping goals into roles and then assigning those roles to agents or teams. The grouping of goals into roles encourages one to think about the types of functionality and responsibility that go together.

The Prometheus methodology [49] recommends designing agent-based systems so that the *coupling* between agents is low, but the *cohesion* of functionality within an agent is high. Low coupling means that each agent is relatively independent, and so changes to an agent's behaviour can be made locally within the agent, without having to change the code of other agents in the system. High cohesion means that the goals of the agent are related to one another in some way, for example, because they involve a similar skill set, use the same belief sets, or involve behaviours that tend to co-occur in time/place. The analysis of coupling and cohesion is a useful heuristic for deciding how to decompose a system into individual agents. However, in the typical scenarios where TDF is applied, we are usually modelling decision-making entities in the real world, and so the problem of how to partition the system into agents is moot; we know what types of agent we need to create. Nevertheless, it is still useful to group goals into roles, particularly where teams are concerned, because it is intuitive to understand a team in terms of the role-playing capabilities of its members.

Although the case study diagram embodies a particular view of the entities involved in the example, one is not compelled to adopt the same perspective when modelling. For example, the current case study conceptualises the submarine as a single decision-making entity – effectively the submarine commander; there is no reference to other decision makers on board the submarine. However, other perspectives are possible; one could distribute the decision-making across a number of agents/teams (and roles) within the submarine. A submarine contains a crew, and each crew member or team has a particular role to play, as implied by the I/O Diagram in Figure 3.2. Typically, the navigation, fire control, sonar and TMA stations are manned by separate teams who enact the roles specific to the functions of those stations. It would be worth capturing these aspects of the internal crew structure of the submarine, if it served the requirements of the modelling objective; for example, if one wanted to run tactical simulations to determine the most efficient crew assignments and how best to roster them. In this case, the `Classify` goal in the strategy in Figure 6.10 could be assigned to a `Sonar Operator` role and the `Track target` goal in Figure 6.11 could be assigned to a `TMA` (Target Motion Analysis) role.

In this tutorial example, the modelling objective is concerned with evaluating the effectiveness of different STG configurations; the submarine decision making only needs to be modelled to the extent necessary for the submarine to react in a tactically appropriate manner during the simulation runs. Therefore, it suffices to model it as a single entity, the `Submarine` agent, and there is no need to structure the agent's functionality and responsibility into roles, although one is free to do so if one feels that adding roles to the `Submarine` agent would clarify what it can do. We have chosen not to partition the `Submarine` agent into roles, and so the goals in the submarine's strategy diagrams (Figures 6.10 to 6.13) will be assigned to the submarine agent later on, during the Behaviour stage.

Turning to the STG navigation strategy (Figure 6.14), and based on the structure of the case study from which this strategy is derived, we can discern the two main roles shown in Figure 6.17. Note that system functionality and responsibilities can be partitioned into roles in a number of ways, and this can sometimes be as much an art as a science. A role can be created for a number of reasons, such as (i) it will involve a particular skill set; (ii) we know in advance that a particular agent/team will have that responsibility; (iii) it helps in partitioning the system's capability in a way that will help others understand its logical structure; or (iv) it is required in order to define the structure of a team in a flexible manner that captures the required skills of the team's members. For example, an alternate role structure could add a `Sonar handling` role to the `Escorts` team, which would then takeover that functionality from the `Threat detection` role[6].

Figure 6.17 shows the role enactments for this tutorial example. Based on the strategy to defend against submarine attack (Figure 6.16), the `Shielding` role has the goal `Shield HVU` and the `Interception` role has the goal `Intercept submarine`. Although these roles are shown now in anticipation of their being used to define the relevant teams, they could have been added later, when the team structure diagram is being created. The TDF methodological process doesn't have to be linear; in general, you will find yourself jumping back and forth between diagrams, as new information comes to light.

---

[6] Sonar handling currently happens within the sector patrol team plan (Figure 6.27), which is triggered by the `Begin patrolling sector` goal that gets delegated to the `Threat detection` role.

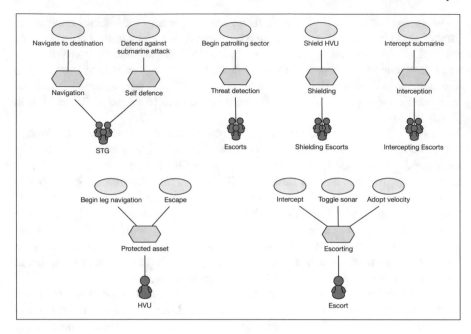

Fig. 6.17 Role enactments

6.4.2 STG Team Structure Diagram

The team structure diagram specifies the team hierarchy that results from adopting a particular goal. The team hierarchy is defined in terms of the roles that the team members need to fill to achieve the overall team goal. Figure 6.18 shows two possible team structures that are formed to handle the 'Navigate to destination' goal (this goal is the root of the STG navigation strategy shown in Figure 6.14). In the case where there is no likely threat (lefthand team structure), the 'STG' team requires two role fillers and has no sub-teams. However, if a threat is likely (righthand team structure), a more complex team is formed with four 'Escorting' role fillers within the 'Escorts' team. In both cases, the team will get disbanded if the 'Navigate to destination' goal is dropped, which happens when there is a submarine threat (as specified in the strategy in Figure 6.14).

The team that is formed to handle the 'Defend against submarine attack' goal adds a 'Shielding' and 'Intercepting' role (Figure 6.19). The 'Intercepting Escorts' sub-team requires two 'Escorting' role fillers, as does the 'Shielding Escorts' sub-team. This team structure definition results in an 'STG' team comprising an 'HVU' agent, an 'Intercepting Escorts' team, as well as a 'Shielding Escorts' team, and each of these sub-teams is made up of two 'Escort' agents.

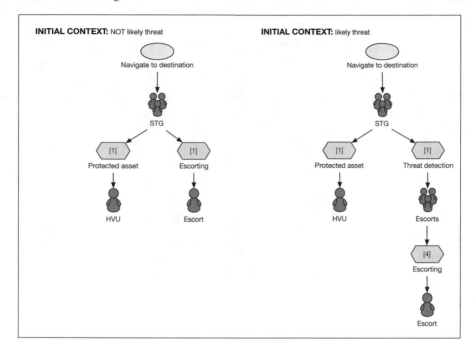

Fig. 6.18 Navigation team structure of STG

6.4.3 Architecture Overview Diagram

Because we will be focusing on the STG team in the next section, the architecture overview diagram (Figure 6.20), focuses on the STG subset of the model. It shows the messaging and actions that are present in the plans shown in Sections 6.5.2 to 6.5.4. The STG navigation team (Figure 6.18) comprises an 'HVU' agent and potentially an 'Escorts' team; the 'Escorts' team in turn is made up of four 'Escort' agents.

The 'Begin patrolling sector' and 'Begin leg navigation' goals derive from the 'Travel for leg duration' team plan (Figure 6.26). The two messages are sent in the 'Sector patrol' team plan (Figure 6.27). Finally, the two HVU actions are performed by the HVU leg navigation plan, shown in Figure 6.29.

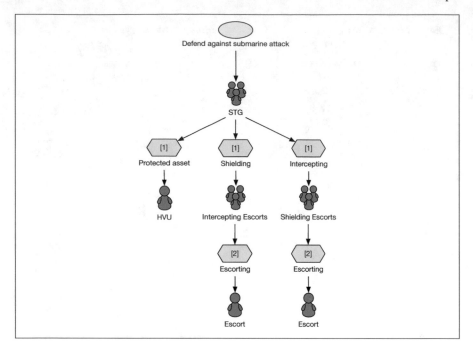

Fig. 6.19 Defensive team structure of STG

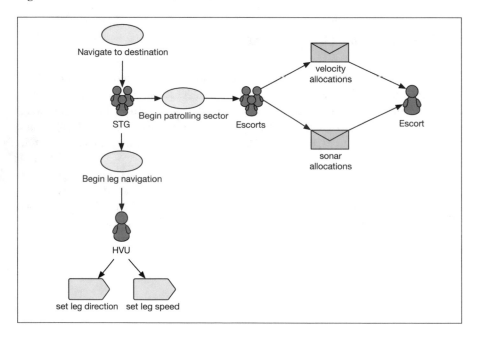

Fig. 6.20 Architecture overview diagram

6.5 Behaviour Stage

Introduced in Chapter 5, the Behaviour stage comprises three main types of agent diagram and also has team-oriented versions of each, making a total of six diagram types. In keeping with its objective of being a flexible methodology that allows the designer to develop a model in their preferred manner, TDF does not mandate a particular order in which to develop the diagrams. Having said that, based on our experience in modelling with TDF, there are some useful heuristics that can expedite the design of agent/team behaviour. The most effective heuristics are listed below, in no particular order.

- **Focus on critical gaps.** If the domain expert's availability is likely to end soon, rather than working on the parts of the design that are well understood, develop those aspects that you feel are unclear. This will help identify important gaps, allowing you to seek clarification from the domain expert before they become unavailable. Otherwise, if you focus on creating tactics and plans for the well understood, more mundane agent/team behaviours, you run the risk of ending up with a trivial model that is unable to handle the more difficult edge cases and tactical scenarios.
- **Focus on what is known.** If domain expert availability is not an issue, develop the parts of the model that you feel are well understood. In this way, you will rapidly map out the behaviour of the agents and teams, and thereby gain an overall understanding that will help with developing the more intricate behaviours of the system.
- **Develop behaviours top-down.** In general, TDF favours a top-down approach to design, and this is reflected in the overall structure of its stages, which move from requirements and knowledge elicitation, through system specification, to the design of the detailed behaviour of the entities that make up the system. In a top-down approach, you should begin with the strategies, roles and team structures developed earlier, and map them to team tactics – effectively the high-level, goal-directed, aggregated behaviour of groups. Then, for each team tactic, design the team plans required to carry out the functionality of the tactic. Finally, focus on the individuals, i.e. the agents, and design them in a similar top-down fashion, from tactics through to plans.
- **Develop behaviours bottom-up.** Sometimes, it is not clear how the behaviour should be structured into tactics, for example, when developing new behaviours in a novel domain. Although such cases are atypical, when they arise, it can be more effective to work bottom-up, i.e. by developing the agent plans and team plans before trying to group them into tactics.

In this tutorial, we will develop the behaviours top-down because the domain is relatively well understood, and furthermore, a top-down approach will provide the clearest exposition of the design. Consequently, we begin by specifying team tactics before drilling down into the detailed team plans; similarly, agent tactics are specified before plans.

6.5.1 STG Team Tactics

In this section, we continue to focus on the behaviour of the STG team. Strategies provide a high-level view of the goal-directed reasoning that occurs in the system being modelled. They serve as both a means of documenting the goal-directed aspects of the design and a way of providing a human team member with an explanation of why the agent or team is behaving as it is. In order to produce behaviour, the strategy must map to tactics and plans. Here we show two tactics that relate to the strategy (Figure 6.14) for handling the 'Navigate to destination' goal. The first tactic (Figure 6.21) maps to the 'Navigate directly' goal; the tactic's objective, 'Navigate directly', is handled by the plan 'Travel directly', and the 'Plan route' plan reads from the 'Mission parameters' belief set and reads from and writes to the 'Route constraints' one. The tactic is able to calculate a leg velocity & duration and then travel for the duration of that leg. Subsequent legs are traversed via recursive adoption of the 'Navigate directly' goal, and will be based upon the 'Route constraints' belief set.

Figure 6.22 presents a tactic for handling the goal 'Navigate defensively'. It has two plans for handling the 'Plan general route' goal; 'Quickly plan route' is chosen if STG needs to depart in less than five minutes from when the goal was adopted, and 'Plan route', which is also used in the other tactic, is used if there is more time to plan how to navigate to the destination. This tactic has the sub-tactic 'Zigzag navigation', which is shown in Figure 6.23. The 'Zigzag navigation' tactic reflects the strategy shown in Figure 6.15 and reveals that the 'Begin leg navigation' and 'Begin patrolling sector' goals are sent to the team members that are enacting the roles 'Protected asset' and 'Threat detection' respectively. This is also reflected in the 'Travel for leg duration' team plan shown in Figure 6.26.

6.5.2 Team Plans

This section presents two STG team plans and one 'Escorts' team plan. The two STG team plans are part of the 'Zigzag navigation' tactic (Figure 6.23). The first plan, 'Zigzag navigation' (Figure 6.24), does not involve any coordination of team members; it handles the 'Zigzag along route' objective of the tactic by sequentially trying to achieve the three sub-goals shown, and has a 'SUCCEED IF: at destination' success condition.

An alternative way of expressing the fact that the plan succeeds once 'at destination' is true, is to add an upper fork branch that comes together via a merge node (Figure 6.25). The merge node means that the plan will complete as soon as either branch does, and so it will keep recursively invoking the 'Zigzag along route' goal until the 'at destination' wait condition becomes true. This procedural encoding of the success condition is more difficult to read, although it is probably closer to how the logic would be implemented on a typical BDI platform.

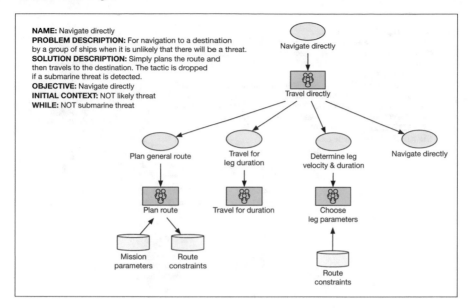

Fig. 6.21 STG tactic to navigate if a threat is unlikely

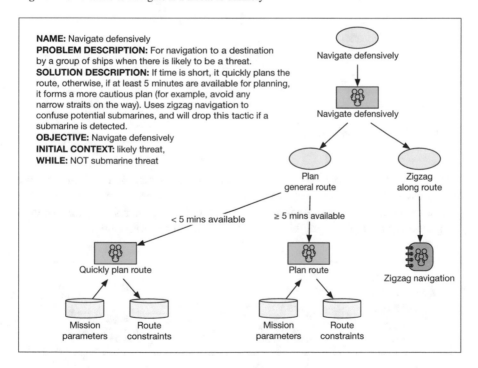

Fig. 6.22 STG tactic to navigate if a threat is likely

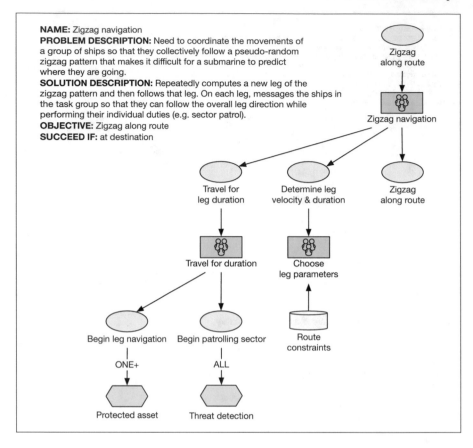

Fig. 6.23 STG tactic to move in a zigzag pattern

The point to take home here is that TDF provides considerable flexibility in how to express a plan-based method of achieving a goal; in general, annotating the plan with a declarative success condition is clearer, but a designer with an imperative programming background might prefer the procedural fork/merge representation.

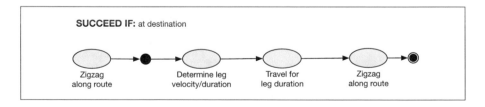

Fig. 6.24 Team plan for zigzag navigation, with declarative success condition

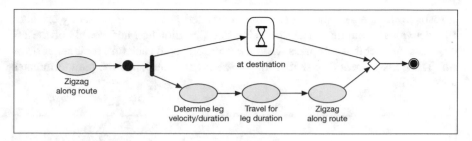

Fig. 6.25 Team plan for zigzag navigation, success condition expressed procedurally

The STG team plan, 'Travel for leg duration', is shown in Figure 6.26. It concurrently sends a 'Begin patrolling sector' goal to the 'Threat detection' role filler and a 'Begin leg navigation' goal to the 'Protected asset' role filler. It then waits for the predetermined duration to elapse, before completing successfully.

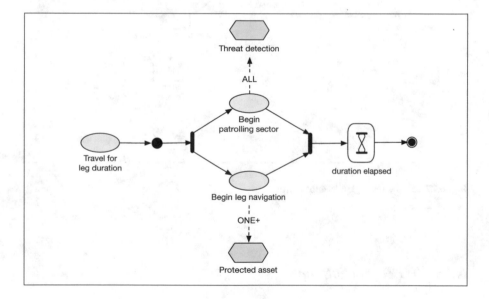

Fig. 6.26 Team plan 'Travel for duration'

The 'Begin patrolling sector' goal, sent from the 'Travel for leg duration' STG team plan, is handled by the 'Escorts' team plan: 'Sector patrol' (Figure 6.27). It delegates a 'Get escort positions' goal to the 'Escorting' role fillers. Depending on which team structure was created during the team formation phase (see Figure 6.18), there will be either one or four 'Escorting' role fillers. Either way, once they have all achieved the goal, the team plan can proceed.

It reads from the `escort positions` belief set[7] and then concurrently executes two activities – one that picks the velocities (i.e. direction and speed) of the escorts and another that computes when the escorts should each turn their sonar on or off. These assignments are sent to the `Escorting` role[8], and the plan terminates successfully.

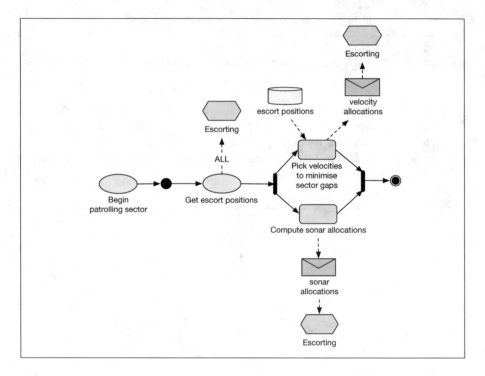

Fig. 6.27 Escorts team plan 'Sector patrol'

6.5.3 STG Team Diagram

The team diagram in Figure 6.28 illustrates the artefacts of the STG team. This diagram is particularly concise because the STG team's functionality has been encapsulated into tactics.

---

[7] Though not shown here due to space constraints, this belief set is shared with the Escorting role fillers by virtue of the fact that the `Escort` agents have a belief set of the same name.

[8] If you want to, for example, send a message to just one of the `Escorting` role fillers, you would have to give it a unique role and then reference that role in the team plan.

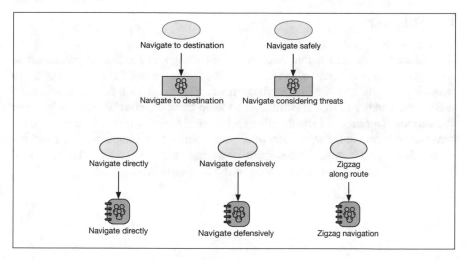

Fig. 6.28 STG team diagram

6.5.4 HVU Navigation Plan

We conclude this tutorial with an HVU agent plan to handle the 'Begin leg navigation' goal that is part of the 'Protected asset' role (Figure 6.17). This plan begins with an activity that chooses the leg duration, direction and speed. It then forks to set the leg speed and direction, before joining the two branches to end the plan.

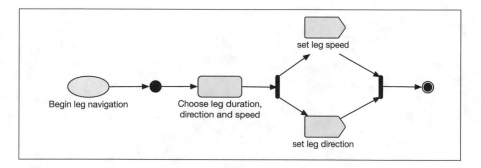

Fig. 6.29 HVU leg navigation plan

6.6 Summary

This chapter presented a thin slice of a full design to handle the `Submarine escapes interception`' case study, and it focused on the STG team in order to provide a feel for how the various diagrams relate to one another. When designing a decision-making model, it is difficult to keep track of the relationships between the various diagrams, and whether these relationships are fully specified and consistent with one another. Propagation of relationships across diagrams, and checking of consistency, is best performed in software rather than manually; the TDF Tool provides this support and is introduced in the next chapter.

Chapter 7
TDF Tool

Drawing TDF diagrams typically involves selecting a diagram type, adding entities to the diagram and then connecting them together. This is time consuming and error prone, whether using pen and paper or a general-purpose drawing application. Not only must one draw the appropriate icons, but one must remember what icons are allowed in each diagram type and the rules for connecting the icons together. Furthermore, the overall model must satisfy a number of implicit constraints; for example, if a team is defined as being able to achieve a particular goal, it must have at least one team plan to handle that goal. For anything but a trivial model, tool support is required in order to automate such consistency checking. The TDF Tool provides this functionality, supporting the creation of the various diagram types, consistency checking and the generation of code templates. It is a free-to-use standalone application that can be downloaded from `http://tdfagents.com`, and it runs on Linux, macOS and Microsoft Windows. In this chapter, we provide a brief snapshot of the TDF Tool's user interface, and then outline the tool's core capabilities, namely consistency checking and code generation.

In the top-left pane of Figure 7.1, the tool reflects the three stages of the TDF methodological process. Each stage can be expanded to provide access to the diagrams it comprises; the Requirements stage is shown expanded here, with a strategy displayed. Using the tool, one can load/save TDF projects, as well as print diagrams.

© Springer Nature Switzerland AG 2019
R. Evertsz et al., *Practical Modelling of Dynamic Decision Making*, SpringerBriefs in Intelligent Systems,
https://doi.org/10.1007/978-3-319-95195-9_7

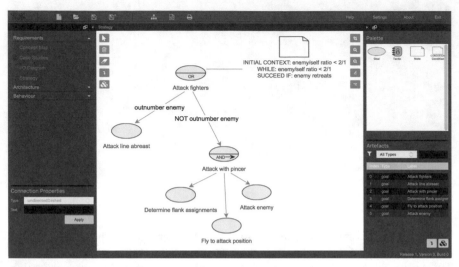

Fig. 7.1 Editing a strategy diagram in the TDF Tool

7.1 Consistency Checking

Although the tool is updated and extended regularly, at a minimum, it provides the core consistency-checking functionality outlined in this section. In the tool, type safety is provided by restricting which artefacts can appear in a diagram and how the artefacts can be linked together. This eliminates a significant source of potential modelling errors. In addition, the tool offers the following consistency-checking functionality:

- **Goal handling.** All goals need to be handled by at least one agent/team plan or tactic. Also, all agent/teams must be able to handle the goals that make up the roles they are declared to enact.
- **Message handling.** All sent messages must be handled by at least one agent/team plan.
- **Architecture overview diagram.** The artefact relationships added to this diagram must be a subset of the propagated[1] ones.
- **Role enactment diagram.** Each role enactor must have a plan for every goal in its role.
- **Team structure diagram.** The team at the root of the team structure must be able to handle its declared team goal by having a plan or tactic that handles it. Also, every team member must be capable of enacting its role(s) in the team.

---

[1] The TDF Tool offers automatic propagation, from one diagram to another, of relationships that hold between the artefacts in the model. So, for example, if the architecture overview diagram shows 'Agent A' as having 'Action B', then that agent must have at least one plan that performs this action. In other words, 'Action B' must be one of the artefact relationships propagated from the list of the agent's plans.

- **Agent and team diagrams.** The artefact relationships added to this diagram must be a subset of the propagated ones.
- **Plan diagram.** If the plan sends a message or delegates a goal, the recipient must have a plan to handle it.
- **Team plan diagram.** Any role referenced in a team plan must be in at least one of the team structures defined for that team. This is because a team plan can only be applicable if all of its referenced roles are present in the team. Therefore, if a team plan references a role that is in none of its possible team structures, it will never be selected, and so is redundant.
- **Tactic diagram.** The tactic must have a plan or sub-tactic that handles the tactic's objective (goal). The artefact relationships added to this diagram must be a subset of the propagated ones.
- **Team tactic diagram.** The team tactic must have a plan or sub-tactic that handles the tactic's objective (goal). The artefact relationships added to this diagram must be a subset of the propagated ones.

In future versions of the TDF Tool, we will be investigating more complex consistency checks and will also look at incorporating techniques for the verification of design models such as those presented in [1, 3, 2, 69, 70].

7.2 Code Generation

Another significant problem for models of decision making has been the fact that they are usually tied to a specific implementation platform, and transferring the models to another platform entails labour-intensive reimplementation. To some extent, TDF has the potential to ameliorate this problem because, as an implementation-independent representation of the decision making of agents and teams, there is the potential to generate platform-specific *code templates* from the designs. Without burdening TDF's design language with more formal code-related annotations, it is not possible to generate a fully running model direct from the design. For example, an activity in a TDF plan has an associated natural language description, but the individual lines of code required to implement the intended semantics need to be written by a programmer who is familiar with the target platform.

Currently, TDF generates JACK code templates from TDF designs, but we plan to extend the code generation capability to encompass other platforms, as and when the need arises. The generated code templates contain annotations from the design that demarcate areas that can be filled out. The code that is added to those areas is preserved by TDF, ready for the next code generation cycle. For example, if the programmer fills out an activity, that code will remain associated with the activity. If the activity is deleted from the design, the designer/programmer is warned that the associated code will be lost.

Chapter 8
Concluding Remarks

This book presented the motivation for, and a methodological approach to, modelling dynamic decision making – or what are generally referred to as *tactics*. The key attribute of tactics is that they are *goal-directed* and also *responsive* to changes in the situation faced by the decision maker. In providing a methodology for modelling tactics, we not only focused on how to express the purposeful aspects of behaviour, but also on how an agent or team's purpose and approach can be adapted to environmental variation.

Another key driver was the need to offer a representation that can be understood by domain experts and human team members. Domain expert understanding of models is vital to tactics development and validation and, in the past, validation has been a significant problem for the developers of dynamic decision-making systems. Recent technological advances in the field of autonomous systems mean that humans will need to work in concert with artificial decision-making systems out in the real world, which is typically a dynamic, fluid environment. Effective joint activity depends on mutual understanding of what the team members are doing and why. Thus, *agent transparency* is a key requirement in human/agent teams. TDF's approach to understandability and transparency is to offer a diagrammatic representation that is based on a good *folk psychology* model of human reasoning; namely, the intentional stance, as conceptualised in the BDI paradigm. In the BDI paradigm, an agent's decision making is understood in terms of its **beliefs** about the world (its situation awareness), what it **desires** to achieve (its goals), and its **intentions** (the goals it is *currently* committed to, and the methods it has decided to use to achieve them).

Based on over 25 years of experience in modelling agents and teams, we distilled a number of key requirements for a methodology and tool that will facilitate the modelling of dynamic decision making. These were outlined at the beginning of Chapter 2 and covered tactics, roles, teams, knowledge elicitation, goal structures, reusable software components, agents, situation awareness, environmental interaction, procedural reasoning, proactive/reactive behaviour, context sensitivity, and explainable agent behaviour; we also noted which aspects of TDF address each requirement.

© Springer Nature Switzerland AG 2019
R. Evertsz et al., *Practical Modelling of Dynamic Decision Making*, SpringerBriefs in Intelligent Systems,
https://doi.org/10.1007/978-3-319-95195-9_8

As a modelling methodology, TDF is unique in addressing all of these requirements. It does so by providing: (i) a *process* for eliciting, modelling and implementing tactical decision-making systems, (ii) a key set of artefacts and relationships that
are important for the effective modelling of dynamic decision-making behaviour,
(iii) a set of diagrams that collect those artefacts and their interrelationships together
into meaningful views of the underlying tactical system, and (iv) support for consistency checking, propagation of implicit relationships, and the generation of code
templates that can support the implementation of agents and teams in software.

Although we believe that TDF offers a fairly comprehensive approach to the
modelling of dynamic decision making, there is of course always room for further
development and improvement. Having said that, a key driver for TDF was the need
for it to be easy to learn and apply. During various phases of its five years of development and evaluation, which in turn was founded on 15 years of experience with
the Prometheus methodology, it was at times a lot more complex than it is now. For
example, the case study notation was originally based on BPMN [48], the de facto
standard for process notation. We spent a considerable amount of time investigating
the utility of BPMN, in line with the following two objectives: (i) the representation should be simple and clear to domain experts, and (ii) it should be capable of
supporting the formal verification of the consistency between a case study and the
resulting TDF model. Objective (i) was tackled by adopting a greatly reduced subset of BPMN notation, and confidence in (ii) came from work on mapping BPMN
to BDI, for example [20]. Despite a number of iterations of simplification, it became clear that domain experts find BPMN to be unintuitive and difficult to learn.
During this period, we also investigated the use of BPMN to represent TDF plan
diagrams, but this too was found to be difficult for users to understand. An earlier
comparison of TDF plan diagrams with UML suggested that TDF plan diagrams
are significantly easier to understand, even for competent UML users who have had
no previous exposure to TDF [27]. Consequently, we adapted TDF's plan diagram
representation so that it could be used for case studies by excluding irrelevant node
types and providing a means of representing communication between actors in the
case study.

With the proviso that TDF should not become too complex to learn and apply,
there are a couple of important areas for further research and development:

- **Explicit traceability between knowledge elicitation and design.** This would
  involve extending the methodology to incorporate a phase in which the modeller
  explicitly maps case study elements to TDF artefacts, for example, identifying
  case study activities as being achieved by particular goals in the model. Tool support could then be provided for verifying that one of the execution paths through
  the model will produce that sequence of goals, and therefore that same sequence
  of activities. Currently, the TDF methodology encourages identifying TDF artefacts (Section 3.2.1.1), such as goals adopted or dropped, but these are annotated
  by **note** nodes, and there is no formal way to relate them to design artefacts.
- **Meta-level reasoning.** A key capability of effective decision makers is choosing
  the appropriate course of action for a given situation, and switching the approach
  if the situation changes or new, clarifying information comes to light. TDF sup-

ports the representation of such dynamic factors by providing goal properties as well as guards on arcs (Figure 2.9). TDF's approach is essentially one of labelling *specific* choices with contextual information. However, a more powerful method of deliberating about a set of options is to reason at a level that is above the concrete situation, namely at the level of the abstract reasoning artefacts themselves. This is termed **meta-level reasoning**. So, rather than directly encoding the choices between options, allow the decision maker to reason about the options themselves and their properties. This could include explicit handling of goal priorities and conflicts, and it could also incorporate the notion of policies and norms. Policies and norms represent the general social understanding that we have about what is appropriate or permissible to do in a given situation. For example, in military tactics, **ROE** (Rules of Engagement) are used to constrain the available tactical options [22]; hence, firing a tank shell at a sniper in a hospital window would generally be excluded by ROE, due to factors such as needing to avoid civilian casualties. The incorporation of meta-level reasoning into a diagrammatic language such as TDF would be challenging, but it is a worthwhile area for further research.

This concludes our exposition of the modelling of dynamic decision making. TDF is intended to be a very practical methodology for specifying, designing and implementing dynamic decision-making systems. We wish you every success in your tactics modelling endeavours!

References

1. Yoosef Abushark, John Thangarajah, James Harland, and Tim Miller. Checking consistency of agent designs against interaction protocols for early-phase defect location. In *Proceedings of AAMAS-2014*, pages 933–940, Paris, France, May 2014.
2. Yoosef Abushark, John Thangarajah, Tim Miller, and James Harland. A framework for automatically ensuring the conformance of agent designs. *Journal of Systems and Software*, 113:266–310, 2017.
3. Yoosef Abushark, John Thangarajah, Tim Miller, James Harland, and Michael Winikoff. Early detection of design faults relative to requirement specifications in agent-based models. In *Proceedings of AAMAS-2015*, pages 1071–1079, Istanbul, Turkey, May 2015.
4. Fahrettin Akbori. Autonomous-agent based simulation of anti-submarine warfare operations with the goal of protecting a high value unit. Master's thesis, 2004.
5. Bernhard Bauer, Jörg P Müller, and James Odell. Agent UML: A formalism for specifying multiagent software systems. *International journal of software engineering and knowledge engineering*, 11(03):207–230, 2001.
6. Jeffrey M. Bradshaw, Paul J. Feltovich, and Matthew Johnson. Human-agent interaction. *The Handbook of Human-Machine Interaction: A Human-Centered Design Approach*, 2012.
7. M. E. Bratman. *Intention, Plans, and Practical Reasoning*. Harvard University Press, Cambridge, MA (USA), 1987.
8. Michael Bratman. *Faces of intention: Selected essays on intention and agency*. Cambridge University Press, 1999.
9. Berndt Brehmer. Dynamic decision making: Human control of complex systems. *Acta psychologica*, 81(3):211–241, 1992.
10. Janet E. Burge. Knowledge elicitation for design task sequencing knowledge. Master's thesis, Worcester Polytechnic Institute, 1998.
11. Balakrishnan Chandrasekaran. Generic tasks in knowledge-based reasoning: High-level building blocks for expert system design. *IEEE expert*, 1(3):23–30, 1986.
12. William G. Chase and Herbert A. Simon. Perception in chess. *Cognitive psychology*, 4(1):55–81, 1973.
13. Jessie Y.C. Chen and Michael J. Barnes. Agent transparency for human-agent teaming effectiveness. In *Systems, Man, and Cybernetics (SMC), 2015 IEEE International Conference on*, pages 1381–1385. IEEE, 2015.
14. Philip R. Cohen and Hector J. Levesque. Teamwork. *Nous*, 25(4):487–512, 1991.
15. Pedro F. Coll. Target motion analysis from a diesel submarine's perspective. Technical report, DTIC Document, 1994.

© Springer Nature Switzerland AG 2019
R. Evertsz et al., *Practical Modelling of Dynamic Decision Making*, SpringerBriefs in Intelligent Systems,
https://doi.org/10.1007/978-3-319-95195-9

16. Robert Coram. *Boyd: The Fighter Pilot Who Changed the Art of War*. Back Bay Books/Little, Brown and Company, New York, NY, 2002.
17. Beth Crandall, Gary A. Klein, and Robert R. Hoffman. *Working minds: A practitioner's guide to cognitive task analysis*. MIT Press, 2006.
18. Lloyd M. Crumley and Mitchell B. Sherman. Review of command and control models and theory. Technical report, DTIC Document, 1990.
19. D. C. Dennett. *The Intentional Stance*. MIT Press, 1987.
20. Holger Endert, Tobias Küster, Benjamin Hirsch, and Sahin Albayrak. Mapping BPMN to agents: An analysis. In *First International Workshop on Agents, Web-Services, and Ontologies Integrated Methodologies*, pages 43–58. MALLOW, 2007.
21. M. R. Endsley. Design and evaluation for situational awareness enhancement. In *Proceedings of the Human Factors Society 32nd Annual Meeting*, pages 97–101, Santa Monica, CA, 1988. Human Factors Society.
22. R. Evertsz, F. E. Ritter, S. Russell, and D. Shepherdson. Modeling rules of engagement in computer generated forces. In *Proceedings of the 16th Conference on Behavior Representation in Modeling and Simulation*, pages 123–134, Orlando, FL, 2007. U. of Central Florida.
23. Rick Evertsz, Martyn Fletcher, Richard Jones, Jacquie Jarvis, James Brusey, and Sandy Dance. Implementing industrial multi-agent systems using JACK. In *International Workshop on Programming Multi-Agent Systems*, pages 18–48. Springer, 2003.
24. Rick Evertsz, John Thangarajah, and Thanh Ly. A BDI-based methodology for eliciting tactical decision-making expertise. *Data and Decision Sciences in Action: Proceedings of the Australian Society for Operations Research Conference 2016*, pages 13–26, 2018.
25. Rick Evertsz, John Thangarajah, and Michael Papasimeon. The conceptual modelling of dynamic teams for autonomous systems. In Heinrich C. Mayr, Giancarlo Guizzardi, Hui Ma, and Oscar Pastor, editors, *Conceptual Modeling*, pages 311–324, Cham, 2017. Springer International Publishing.
26. Rick Evertsz, John Thangarajah, Nitin Yadav, and Thanh Ly. Agent oriented modelling of tactical decision making. In Pinar Yolum, Gerhard Weiss, Edith Elkind, and Rafael H. Bordini, editors, *Proceedings of the 14th International Conference on Autonomous Agents and Multiagent Systems*, pages 1051–1060, Istanbul, 2015.
27. Rick Evertsz, John Thangarajah, Nitin Yadav, and Thanh Ly. A framework for modelling tactical decision-making in autonomous systems. *J. Syst. Softw.*, 110(C):222–238, 12 2015.
28. Edward A. Feigenbaum. Themes and case studies of knowledge engineering. *Expert Systems in the Micro-Electronic Age*, pages 3–25, 1979.
29. Paul J. Feltovich, Jeffrey M. Bradshaw, William J. Clancey, and Matthew Johnson. Toward an ontology of regulation: Socially-based support for coordination in human and machine joint activity. In *International Workshop on Engineering Societies in the Agents World*, pages 175–192. Springer, 2006.
30. Erich Gamma, Richard Helm, Ralph Johnson, and John Vlissides. *Design patterns: Elements of reusable object-oriented design*. Addison-Wesley Reading, 1995.
31. M. P. Georgeff and A. L. Lansky. Development of an expert system for representing procedural knowledge. Final report, for NASA Ames Research Center, Moffett Field, California, USA. *Artificial Intelligence Center, SRI International, Menlo Park, California, USA*, 1985.
32. Frank Bunker Gilbreth and Lillian Moller Gilbreth. Process charts - first steps in finding the one best way to do work. Paper presented to the American Society of Mechanical Engineers, 1921.
33. Wayne D. Gray, Susan S. Kirschenbaum, and Brian D. Ehret. The précis of Project Nemo, phase 1: Subgoaling and subschemas for submariners. In *Nineteenth Annual Conference of the Cognitive Science Society*, pages 283–288, 1997.
34. James Harland, David N. Morley, John Thangarajah, and Neil Yorke-Smith. Aborting, suspending, and resuming goals and plans in BDI agents. *Autonomous Agents and Multi-Agent Systems*, 31(2):288–331, 2017.
35. Linda Heaton. Unified modeling language (UML): Superstructure specification, v2.0. *Object Management Group, Tech. Rep*, 2005.

36. Clint Heinze, Michael Papasimeon, and Simon Goss. Issues in modelling sensor and data fusion in agent based simulation of air operations. In *Proc. 6th Int. Conf. on Information Fusion*, 2003.
37. François Félix Ingrand, Raja Chatila, Rachid Alami, and Frédérick Robert. PRS: A high level supervision and control language for autonomous mobile robots. In *Robotics and Automation, 1996. Proceedings, 1996 IEEE International Conference*, volume 1, pages 43–49. IEEE, 1996.
38. W. Lewis Johnson. Agents that learn to explain themselves. In *AAAI*, pages 1257–1263, 1994.
39. Gary A. Klein. *Sources of power: How people make decisions*. MIT Press, Cambridge, MA, 1998.
40. Joel Lawson. Command control as a process. *IEEE Control Systems Magazine*, 1(1):5–11, 1981.
41. John D. Lee and Katrina A. See. Trust in automation: Designing for appropriate reliance. *Human factors*, 46(1):50–80, 2004.
42. Brian Logan, John Thangarajah, and Neil Yorke-Smith. Progressing intention progression: A call for a goal-plan tree contest. In S. Das, E. Durfee, K. Larson, and M. Winikoff, editors, *Proceedings of the 16th International Conference on Autonomous Agents and Multiagent Systems (AAMAS 2017)*, pages 768–772, Sao Paulo, Brazil, May 2017. IFAAMAS.
43. John McDermott. Preliminary steps toward a taxonomy of problem-solving methods. In *Automating knowledge acquisition for expert systems*, pages 225–256. Springer, 1988.
44. Joseph E. Mercado, Michael A Rupp, Jessie Y.C. Chen, Michael J. Barnes, Daniel Barber, and Katelyn Procci. Intelligent agent transparency in human–agent teaming for multi-UxV management. *Human factors*, 58(3):401–415, 2016.
45. Bonnie M Muir. Trust between humans and machines, and the design of decision aids. *International Journal of Man-Machine Studies*, 27(5-6):527–539, 1987.
46. G. Murray, D. Steuart, D. Appla, D. McIlroy, C. Heinze, M. Cross, A. Chandran, R. Raszka, G. Tidhar, A. Rao, A. Pegler, D. Morley, and P. Busetta. The challenge of whole air mission modelling. In *Proceedings of the Australian Joint Conference on Artificial Intelligence*, Melbourne, Australia, 1995.
47. Emma Norling. Folk psychology for human modelling: Extending the BDI paradigm. In *Proceedings of the Third International Joint Conference on Autonomous Agents and Multiagent Systems-Volume 1*, pages 202–209. IEEE Computer Society, 2004.
48. OMG. Business process model and notation (BPMN) - OMG Document Number: formal/2011-01-03, 2011.
49. L. Padgham and M. Winikoff. *Developing intelligent agent systems: a practical guide*. Wiley, 2004.
50. Carl A. Petri. Communication with automata: Volume 1 supplement 1. Technical report, DTIC Document, 1966.
51. Robert B. Polk. A critique of the Boyd theory - is it relevant to the army? *Defense Analysis*, 16(3):257–276, 2000.
52. A. S. Rao and M. P. Georgeff. BDI agents: From theory to practice. In *Proceedings of the first ICMAS (95)*, pages 312–319. San Francisco, 1995.
53. Axel Schulte, Diana Donath, and Douglas S. Lange. Design patterns for human-cognitive agent teaming. In *International Conference on Engineering Psychology and Cognitive Ergonomics*, pages 231–243. Springer, 2016.
54. John R. Searle. Responses to critics of the construction of social reality. *Philosophy and Phenomenological Research*, 57(2):449–458, 1997.
55. Nigel Shadbolt and Paul R. Smart. Knowledge elicitation: Methods, tools and techniques. In J. R. Wilson and S. Sharples, editors, *Evaluation of Human Work (4th ed.)*. CRC Press, Florida, USA, 2015.
56. B. G. Silverman. *Human performance simulation*, pages 469–498. Elsevier, Amsterdam, 2004.
57. Glenn Taylor, Randolph M. Jones, Michael Goldstein, Richard Frederiksen, and Robert E. Wray III. VISTA: A generic toolkit for visualizing agent behavior. *Ann Arbor*, 1001:48105, 2002.

58. Glenn Taylor and Robert E. Wray. Behavior design patterns: Engineering human behavior models. In *Proceedings of the Behavior Representation in Modeling and Simulation Conference*, 2004.
59. John Thangarajah, James Harland, David Morley, and Neil Yorke-Smith. Suspending and resuming tasks in BDI agents. In *AAMAS '08: Proceedings of the 7th International Conference on Autonomous Agents and Multiagent Systems*, pages 405–412, Estoril, Portugal, 2008.
60. John Thangarajah, Lin Padgham, and Michael Winikoff. Detecting and avoiding interference between goals in intelligent agents. In *IJCAI '03: Proceedings of the International Joint Conference on Artificial Intelligence*, pages 721–726, Acapulco, Mexico, 2003.
61. John Thangarajah, Lin Padgham, and Michael Winikoff. Detecting and exploiting positive goal interaction in intelligent agents. In *AAMAS '03: Proceedings of the 2nd International Conference on Autonomous Agents and Multiagent Systems*, pages 401–408, Melbourne, Australia, 2003.
62. John Thangarajah, Michael Winikoff, Lin Padgham, and Klaus Fischer. Avoiding resource conflicts in intelligent agents. In *ECAI '02: Proceedings of the European Conference on Artificial Intelligence*, pages 18–22, Lyon, France, 2002.
63. U.S. Army Training and Doctrine Command. Force XXI operations: A concept for the evolution of full-dimensional operations for the strategic army of the early twenty-first century. *TRADOC Pamphlet 525–5*, 1994.
64. Michael Van Lent, William Fisher, and Michael Mancuso. An explainable artificial intelligence system for small-unit tactical behavior. In *Proceedings of the National Conference on Artificial Intelligence*, pages 900–907. Menlo Park, CA; Cambridge, MA; London; AAAI Press; MIT Press; 1999, 2004.
65. Arlette van Wissen, Ya'akov Gal, B.A. Kamphorst, and M.V. Dignum. Human–agent teamwork in dynamic environments. *Computers in Human Behavior*, 28(1):23–33, 2012.
66. Kurt VanLehn. Problem solving and cognitive skill acquisition. In M. Posner, editor, *Foundations of cognitive science*, pages 526–79. MIT Press, MA, 1989.
67. Joseph G. Wohl. Force management decision requirements for air force tactical command and control. *IEEE Transactions on Systems, Man, and Cybernetics*, 11(9):618–639, 1981.
68. Michael Wooldridge. *An introduction to multiagent systems*. Wiley, 2008.
69. Nitin Yadav and John Thangarajah. Checking the conformance of requirements in agent designs using ATL. In G. A. Kaminka, M. Fox, P. Bouquet, Hullermeijer E., Dignum F., Dignum V., and van Harmelen F., editors, *Proceedings of ECAI-2016*, pages 243–251, The Hague, The Netherlands, August 2016. ECCAI, IOS Press.
70. Nitin Yadav, John Thangarajah, and Sebastian Sardina. Agent design consistency checking via planning. In *International Joint Conference on Artificial Intelligence (IJCAI)*, pages 458–464, 2017.
71. Yuan Yao, Brian Logan, and John Thangarajah. Intention selection with deadlines. In G. A. Kaminka, M. Fox, P. Bouquet, Hullermeijer E., Dignum F., Dignum V., and van Harmelen F., editors, *Proceedings of the 22nd European Conference on Artificial Intelligence (ECAI-2016)*, pages 2558–2565, The Hague, The Netherlands, August 2016. ECCAI, IOS Press.
72. Yuan Yao, Brian Logan, and John Thangarajah. Robust execution of BDI agent programs by exploiting synergies between intentions. In *Proceedings of the Thirtieth AAAI Conference on Artificial Intelligence, February 12-17, 2016, Phoenix, Arizona, USA*, pages 2558–2565, 2016.
73. Andrew S. Young and Karen A. Harper. TRACE: An ontology-based approach to generic traceability tools for human behavior models. In *Proceedings of the 14th Conference on Behavior Representation in Modeling and Simulation (BRIMS05)*, 2005.

Printed in the United States
By Bookmasters